Numanities - Arts and Humanities in Progress

Volume 4

Series editor

Dario Martinelli, Kaunas, Lithuania

The series originates from the need to create a more proactive platform in the form of monographs and edited volumes in thematic collections, to discuss the current crisis of the humanities and its possible solutions, in a spirit that should be both critical and self-critical.

"Numanities" (New Humanities) aim to unify the various approaches and potentials of the humanities in the context, dynamics and problems of current societies, and in the attempt to overcome the crisis.

The series is intended to target an academic audience interested in the following areas:

- Traditional fields of humanities whose research paths are focused on issues of current concern;
- New fields of humanities emerged to meet the demands of societal changes;
- Multi/Inter/Cross/Transdisciplinary dialogues between humanities and social and/or natural sciences;
- Humanities "in disguise", that is, those fields (currently belonging to other spheres), that remain rooted in a humanistic vision of the world;
- Forms of investigations and reflections, in which the humanities monitor and critically assess their scientific status and social condition;
- Forms of research animated by creative and innovative humanities-based approaches;
- Applied humanities.

More information about this series at http://www.springer.com/series/14105

Roberto Marchesini

Over the Human

Post-humanism and the Concept of Animal Epiphany

Springer

Roberto Marchesini
Study Centre of Posthuman Philosophy
Bologna
Italy

Translated by Sarah De Sanctis

ISSN 2510-442X ISSN 2510-4438 (electronic)
Numanities - Arts and Humanities in Progress
ISBN 978-3-319-62580-5 ISBN 978-3-319-62581-2 (eBook)
DOI 10.1007/978-3-319-62581-2

Library of Congress Control Number: 2017946921

Printed on acid-free paper

This Springer imprint is published by Springer Nature
The registered company is Springer International Publishing AG
The registered company address is: Gewerbestrasse 11, 6330 Cham, Switzerland

Contents

Chapter 1
The Epimethan Condition

Do animals exist? Or are they only a construction, a polarizing mirror that highlights the excellence and special nature of human beings—as if they were animals purified from every contamination? This question has been asked many times in the past decades and remains unanswerable if we are stuck in the humanist dichotomy treating man and the animal as mutually exclusive. In this perspective, in order to configure the image of the human it is necessary to oppose it to a background, not confusing it in the zoological magma nor letting it be swallowed by the predicative multi-shapedness of biodiversity. The humanistic imperative is therefore not to turn human peculiarity into a predicate—the human as bearer of a certain specialization—as this would nullify what is proper of human beings, blending it into the *mare magnum* of biodiversity. Beings belonging to other species, with their plural characterization, are nullified in the term "animal" if the latter, far from signifying the condition of "animal-being" including humans, is used as opposed to the human.[1]

In this logic there is no more space for the multi-shapedness of the animal condition, where the predicate of biodiversity is essential to its foundation, but there is a homologating categorization that proceeds negatively, in terms of what is lacked, so that the animal is a being *lacking something* compared to the human. To support the anthropocentric project, in fact, it is indispensable that there is a gap between the human and all other animals: it is not sufficient to declare the former's

[1]J. Derrida, *The Animal That Therefore I Am* in *Critical Inquiry*, Vol. 28, No. 2. (Winter, 2002), pp. 369–418. Derrida writes "'the Animal,' as if all nonhuman living things could be grouped without the common sense of this "commonplace," the Animal, whatever the abyssal differences and structural limits that separate, in the very essence of their being, all "animals," a name that we would therefore be advised, to begin with, to keep within quotation marks. Confined within this catch-all concept, within this vast encampment of the animal, in this general singular, within the strict enclosure of this definite article ("the Animal" and not "animals"), as in a virgin forest, a zoo, a hunting or fishing ground, a paddock or an abattoir, a space of domestication, are all the living things that man does not recognize as his fellows, his neighbors, or his brothers" p. 402.

R. Marchesini, *Over the Human*, Numanities - Arts and Humanities in Progress 4, DOI 10.1007/978-3-319-62581-2_1

specificity. This way the non-human becomes a solid and consistent category: all animals are lacking whatever makes the human a non-animal. This nullifies:

(a) on the one hand, the common animal-being inclusive of the species *Homo sapiens*, so that the human finds itself belonging to a different realm, which requires disciplines and approaches opposed to natural sciences;
(b) on the other hand, heterospecificity[2] as such, as belonging to a peculiar domain (the species) irreducible to whatever claim of functionality opposed to the human.

To rediscover animals and "the animal that therefore I am"[3] it is thus necessary to go beyond mere description, because any description of the animal is the outcome of prejudice, to put it with Gadamer.[4] Before looking at heterospecificity as the very condition of animal-being, it is indispensable to understand the mechanisms that have led to the emergence of the "animal" category/dimension as counter-term of the human condition. It's not an easy task, not only due to cultural frameworks—*in primis*, as we shall see, that of humanism—that have grounded their ideology on this antinomy, but also due to the very mechanism of the encounter with otherness. In other words, there is an intrinsic difficulty in recognizing the other species, which I define "animal epiphany." The encounter with the non-human animal is hardly contained within the phenomenic area, which would entail remaining in that limbo of objective recognizability made possible by Heidegger's distancing process.

It is certainly true that, to pass from perceiving-using to the neutral perceiving-evaluating that recognises being in itself, it is necessary to take a distance. However, it is equally true that when intersubjectivity occurs—seeing oneself in the face of the other, as suggested by Lévinas[5]—the other loses its characters of objective extraneity and becomes a for-itself, inaugurating a perceiving-mirroring. In other words, we are faced with a process of decentralization (one that takes distance from being as usable but also from the subject evaluating the in-itself) that goes beyond the objectification of the being. When seeing herself in the face of the

---

[2]By "heterospecificity" I mean the characteristic of belonging to a species other than the human. See glossary.

[3]Ibid.

[4]H.G. Gadamer, "The Problem of Historical Consciousness" in *Graduate Faculty Philosophy Journal*, Volume 5, Issue 1, Fall 1975.

Special H.G. Gadamer Issue, pp. 8–52, Here Gadamer states: "To denounce something as prejudice is to suspend its presumed validity; in fact a prejudice in the strict sense of that term cannot get hold of us unless we are sufficiently unconscious of it" p. 48.

[5]E. Lévinas, *Totality and Infinity. An Essay on Exteriority*. Duquesne University Press, Pittsburgh, Pennsylvania, 2011. That is an otherness that acts as a mirror. For Lévinas, we must respect the Other not because we come into contact with it in many different contexts, but because we attribute a meaning to it or, better, as the philosopher writes: "The face speaks. The manifestation of the face is already discourse," p. 66.

non-human animal, as a reflective mirror of the self, the human is absorbed in the other-by-species and recognizes herself in the amazement at being beyond the givenness of her body.

In the face of otherness the human being designs herself from the point of view of identity by means of the spurious image reflected by the other. This is why I speak of "epiphany"[6]: an annunciation, an appearance of the phenomenon in itself, an alternative path to being blindly focused on oneself. As we will see, in the encounter with the non-human animal, there are simultaneously:

(a) a process of identification. It lies in granting the other individuality—no longer a cat, but *that* cat—or a being-in-the-world that transcends membership and brings the relation to it to the intersubjective dimension (I look at you, you look at me) founded on a common basis, i.e. characters that we share. It also lies in being both included in an individual moment of encounter, in a one to one relationship[7];
(b) a process of distancing. It lies in granting the other its own being—cat-being—which means seeing the world through a different perspective, being amazed at diversity, challenging any narcissistic projection and pushes expressive categories, forcing them to take an alternative route. This, however, does not distance the human from the animal, but the human from itself, capturing it even more deeply and de-centering it.

It is in this double movement, characterized by recognition and disavowal, that epiphany occurs, and the result is anything but neutral, because in return the human being finds itself changed, infected by animal otherness. Only then will the non-human animal become an otherness, only when it is recognized as similar and engaging as well as different. The dialogue to which the encounter with the non-human animal forces us leaves no room for separation: it is the feeling of the understanding and dialectical gaze of the other, capable of bringing out the non-obvious datum of a condition not previously assumed. When the non-human animal stops being a phenomenon, the moment when the human being recognizes itself/the non-human animal, immediately there is a kind of partnership between the animal otherness and the human being reflected in it, creating in a hybrid image that in itself is already able to indicate a path of transformation of the human.

---

[6]By "epiphany" I mean the human projection into otherness. See glossary.

[7]Jacques Derrida, when discussing this character of "individuality," writes: "It is true that I identify it as a male or female cat. But even before that identification, I see it as this irreplaceable living being that one day enters my space [...]. Nothing can ever take away from me the certainty that what we have here is [...] a mortal existence, for from the moment that it has a name, its name survives it. It signs its potential disappearance. Mine also, and this disappearance, from that moment to this, *fortlda*, is announced each time that [...] one of us leaves the room." Derrida, *The Animal That Therefore I Am*, p. 379.

1.1 The Evanescence of Animality

Before analyzing the characteristics of this encounter, as I am speaking to a reader who comes from a specific cultural tradition, I think it is a necessary to look at the hermeneutical frames that make the meeting with the non-human animal difficult, or often impossible, within the humanistic paradigm. The non-human has been reduced in terms of "undifferentiated-animal"—a material easily converted in many different processes, such as:

(a) the transformation into figure, to define an opposition, an emancipation, a regression, a revelation, a stigma;
(b) the reduction to concept, movable and usable outside of a specific and concrete reference to the animal as other-by-species compared to the human being;
(c) the metamorphosis of the animal into a picture, called to represent concepts or entities that are otherwise difficult to configure metaphorically, metonymically, symbolically, allegorically, as double or iconic projection;
(d) the construction of transitional or subrogative entities like the anthropomorphic pet, the animal slave, the mechanical or experimental model, the reified and commodified animal.

In its lack of specificity, of its own inalienable *quid*, the undifferentiated-animal is unable to show anything to the human being, neither in phenomenal terms nor, even less so, in epiphanic ones. When I talk about animal epiphany, in fact, I do not mean a nullification of the phenomenal meaning of the non-human—the reduction following the alienation of the specificity—but an overload in terms of deviation from the expectations and projective reductions. The phenomenon, i.e. the sharpness of the predicative specificity, must not in any way be impeded but, on the contrary, it should be emphasized to produce the epiphany. Only an accentuation of the non-human animal's essence is able to have a morphopoietic[8] effect on my body in the hybridational sense, because it is capable of projecting on my human body its heteromorphia.[9] Let me make an example: the epiphanic encounter with an eagle can only be realized as long as it is not transformed into an undifferentiated being (lacking a specific being-in-the-world) approximate to me—this is anthropomorphism—i.e. into the lowest common denominator between the two of us, which would inevitably annihilate the differences. The epiphany can emerge only if you determine the recognition of a common root that does not impede but, for this very reason, emphasizes the differences to the point of absorbing me in a birdlike perspective that is viable from an anthropopoietic perspective. Thus the perspective of "you can fly" opens up before me.

This is why the contemporary tendency to anthropomorphize animals greatly reduces the epiphanic potential of the encounter: many people live with a dog or a cat but very few are "dogmorphised" or "catmorphised", transforming their

[8]For "morphopoietic" see glossary.

[9]For "heteromorphia" see glossary.

experience into an opportunity of development. To do that it would be necessary to maintain and, I would say, especially value the predicate of species-specific diversity, bringing to the surface and magnifying the features that characterize the peculiarities within the common animal condition. Each species has a precise adaptive characterization, some variation of the condition of animal-being in a given context (the savannah or the forest, the gloomy underground or the bright meadow, symbiosis or autonomy) and in a given style (herbivore or carnivore, sedentary or migratory, daily or nightly). These characteristics are both:

(a) a way of reconstructing reality through the act of perception;
(b) a set of coordinates of reference-fruition of the world on the basis of precise operational schemes—as suggested by von Uexküll with the concept of *Umwelt*.[10]

The plurality of predicates characterizing biodiversity allows for no type of categorization except the omnicomprehensive one of animal-being. In other words, each species is different, unique and superior to all the others for a certain predicate; you could build a pyramid placing it at the top only choosing its excellence as a touchstone. The predicates with which the species declines its performative adaptation are in fact silent in terms of oppositional categorization: they do not allow one to create a category that includes all non-human species in opposition to another category that characterizes the human being. Therefore, if we refer to the predicates expressing the particular specialization of each animal, we cannot build an oppositional dichotomy between humans and other species. To create a gap between the living, it is essential to treat predicates like appearance and irrelevance in the definition of the condition itself. In other words, be it a swallow circling in the sky or a dolphin pirouetting in the water, a mole underground or a monkey on a branch, the condition of those living beings does not change. In fact, it is not the diversity/specificity of their predicates that defines their being in the world, but the rootedness that these predicates produce, in their total performative adequacy, denying freedom to the non-human.

The predicates are appearance and the humanist philosopher must be able to see in the socket, in the wings, in the fins nothing but chains that prevent the non-human from any real presence in the world. But, of course, the non-human animal can resist as an entity only by defending the specialty of its predicates, avoiding the uniformation process that turns them into simple chains. Mostly, the non-human animal can may appear to us only if its predicates of specificity haven't been disposed of. This however is not in the anthropocentric precept whose priority is the definitive separation of man from all other species, sanctioning the inexorable

[10]J. Von Uexküll, *Foray into the Worlds of Animals and Humans: With A Theory of Meaning*, University of Minnesota Press, 2010. The ethologist defines the *Umwelt* (world-environment) as the set of the perceived world and the operational world of each animal species. He writes: "everything a subject perceives belongs to its perception world (*Merkwelt*), and everything it produces to its effect world (*Wirkwelt*). These two worlds, of perception and production of effects, form one closed unite, the *environment*" p. 42.

de-flow of predicates. The differentiation is therefore sought not in the content of diversity of the predicate—the hoof of a horse, the fins of a dolphin or the wings of a bat—but in the character of strict adaptive adherence of the predicate that the non-human animal presents.

To bring out the dichotomy we must make the non-human animal can evanescent, rendering it a term that cannot be said in the singular,[11] i.e. we must delete the hoof, fins and wings and bring out the captivation that they produce—nature's inescapable rootedness in nomotethicity.[12] So, without fins there's no longer a dolphin, without wings there's no longer a bat, without hooves there's no longer a horse, but only the animal that is said in the singular—that can only be said in the singular. That's what emerges freeing itself from the appearance of adaptive form. In this perspective, it becomes useless to get lost in the labyrinths of biodiversity, searching for the condition in the declinative predicate because, for the animal said in the singular, being a category is given by a condition that subsumes all predicates. The animal stripped of its connotations of heterospecificity—with respect to biodiverse-world and not only to humans—has nothing to say to the human being and does not deserve any attention since its dimensional character can be derived negatively by the human. Thus we come to the answer that I have so often heard in philosophical dissertations: namely, that it is not necessary to know the animals to know what they are.

For this reason, making the peculiarity of human beings coincide with their declinative[13] specialization (i.e. their *Umwelt*) cannot render the specialness that they claim to have, compared to non-humans. In fact, every animal is different and unique in their declination of their *Umwelt*: a bearer of precise predicates of relation and effectiveness with respect to the world. Thus, diversity should be sought in the content of adherence: in the heaviness of the declinative predicate. Picture a scale: to lower the animal's plate we must make performative predicates heavier, while to raise the human's plate we must do the opposite and lighten functional predicates. In the classical tradition recovered by humanists, the predicative difference in animals is a gift from Epimetheus[14]: full participation in the world but also

---

[11]For the concept of animal as "captivated entity" see the analysis of animality in Heidegger offered by Giorgio Agamben (*The Open. Man and Animal.* Stanford University Press 2003). He writes: "Heidegger seems here to oscillate between two opposite poles, which in some ways recall the paradoxes of mystical knowledge—or, rather, nonknowledge. On the one hand, captivation is a more spellbinding and intense openness than any kind of human knowledge; on the other, insofar as it is not capable of disconcealing its own disinhibitor, it is closed in a total opacity. Animal captivation and the openness of the world therefore seem related to one another as are negative and positive theology, and their relationship is as ambiguous," p. 59.

[12]For "nomotethicity" see glossary.

[13]For "predicate and declination of predicates" see glossary.

[14]For "Epimetheus" see glossary.

radicalization within a specific dimension. This is what humanism rejects: the definition of a circumscribed range that defines a specialty while limiting human freedom.[15]

However, that "something special" that humanism attributes to mankind lies not in adaptive plurality—it is not part of Epimetheus' gift. On the contrary: it is necessary that Epimetheus forgets about mankind, because the emptier the plate, the higher the human goes. This is why the animal is so much present in philosophical discourse on man while being evanescent and hard to pinpoint in the categorial fading built to make the human emerge. The animal doesn't show itself, it hides in a forest of contrastive predications, hiding in fractalic oppositions and symbolic inclusions. Thus, lost in this labyrinth of dichotomies, in the end we're left with nothing. The impression is that, getting rid of the non-human animal, we lose much of the human too. So what is the definition of animal? That is, is there a definition of animal or is it just an illusion?

The animal is what comes before man—the shared ancestral, the last drip of the openness that would make us fully enjoy the world's here-and-now again—or perhaps it is what man never was: the unknown and unknowable, the polarity that has to remain such. Which is true? Well, when we try to define the human we have to resort to the predicative specification of the animal and this should lead us to conclude that the animal—as the human's opposite—has only a meaning if and when it is introjected in the human. First there is only the non-human animal: that is, the plural devoid of a categorial dimension. Thus, we need to understand what has led to the slow erosion of the non-human animal in philosophical thought. We might suppose that it was because of the human need to emerge compared to other species. This, is turn, may depend on:

1. the mechanisms of identity construction that—due to an interpretative bias—lead to emphasize the difference between man and other species, nullifying the differences between the latter, in a way similar to the concept of "barbarian" in the Hellenic world;
2. the a posteriori justification of animal exploitation that, especially after the Neolithic period, mankind has carried out systematically by alienating many heterospecifics from their habitat and lifestyle, turning them into brutes and nullifying their specificity[16];

[15]If we follow humanist thought, from Pico della Mirandola (man as rank-less) to Martin Heidegger (man as world-creator) we'll see that the difference between man and animal lies not in predicates, but in something prior to those. Man is simply not an animal and cannot be differentiated from non-humans by means of predicates.

[16]Those belonging to a species other than *Homo sapiens* are not discriminated by virtue of an ideology, but due to a historical and social structure of exploitation deriving from post-Neolithic practices and, in particular, the process of domestication. There are still many positions with regard to this debate. See the Italian philosopher Marco Maurizi, *Al di là della natura. Gli animali, il capitale e la libertà*, Novalogos, Aprilia, 2012. See also the collection of essays edited by John Sanbonmatsu,*Critical Theory and Animal Liberation*, Rowman & Littlefield Publishers, Maryland, 2010. Along the same post-Marxist line of thought see also D. NibertNibert, David, *Animal*

3. the consequence of the human appropriation of the non-human's predicate, so that animal epiphany (understood as seeing oneself in the heterospecific dimension) starts a hermeneutic circle that can lead to the heterospecific' loss of predicates, which are taken over by the human.

While deeming the first two explanations plausible and probably mutually exclusive, I believe the third to be the most relevant. When epiphany brings about a therianthropic human being (that is, a human being changed through the dimensional hybridization that takes place in the encounter with the non-human) there is a decentering for which the human can look at itself (by self-distancing) and go beyond the dimension given to it by philogenesis.[17] With animal epiphany the human being encounters new existential dimensions beyond its own. This is the *ubermench* that, imagining himself in a therianthropic way, experiences the bird's flight or the bull's strength while perceiving a distance between his own identity (magmatic and changing) and the animal's (stable in its own predicates). It's as if, through epiphany, man took away the dimensional meaning from the heterospecific's predicates—in particular, he takes away the predicate's ownership, so that it becomes an endowment to be freely used by the human. In this way, the predicates turn into strings moving the animal puppet. Losing its ownership, the heterospecific's predicate changes from endowment of animal subjectivity to imperative or mechanism binding and de-subjectivizing the animal.

The animal, as the expression of a category, cannot be before the human nor, at the same time, can it be like the human. In fact, the moment they meet, there is only the spark of experiential subjectivity—wonder in its possible emotional or cognitive declinations. In other words, there is the emergence of the Heideggerian phenomenon that qualifies the interlocutor as being as such. There is no doubt that, in the passage between the objectification of the non-human animal and the subject's mirroring in the latter, the predicate is both emphasized and appropriated. When I speak of seeing oneself in the non-human animal I mean to underline the main character of the epiphany: it can lead one to wish to assume the non-human animal dimension but also, vice versa, to differ even more from it. Therefore, epiphany always produces an appropriation of heteromorphia[18]: one is led to transform it into an anthropopoietic trace, introjecting it.

The animal—as that with which I need to confront myself—always comes after and beyond the human, when the non-human animal transcends its status of extraneity and, through a path of assimilative reference, is brought to being an entity related to the human. Thus there has to be a moment of metamorphosis able to turn the encounter into a relationship, where the heterospecific' manifestation goes beyond its phenomenic character, loosing its extraneity and becoming

(Footnote 16 continued)

Rights/Human Rights. Entanglements of Oppression and Liberation, Rowman & Littlefield Publishers, Lanham, 2002.

[17]For "philogenesis" see glossary.

[18]For "heteromorphia" see glossary.

something that concerns me. The animal therefore comes after the human and its possibility to look at itself by referring to the animal that has lost its character as such. Thus we must admit that only an epiphanic event can turn the non-human animal into a real otherness, but also that only an animal epiphany can realize the emergence of the human as a place of ontological redefinition.

To understand this relationship, only apparently based on biodiversity, it is necessary to investigate all the relations (trying not to get lost in the course of time) to infer the common places of mutual exchange. Animal epiphany is an important and particular event that has been neglected in human history. It is the moment in which the encounter with the non-human animal begins to interrupt the human being's closure, making it an evolving system. Only then can the human see itself and reflect on itself, thanks to the decentred position it has assumed through the contamination with otherness. If man, as a philogenetic entity, can be evaluated through a simple overview of his inner predicates, the human dimension is always dialogic, because it is the outcome of hybridization with the non-human. Thus animal epiphany, as a spark able to crack open the wall of solipsism of philogenetic man, requires special attention: it is only in its emergent process that a human being can see itself.

This means that the animal that therefore I am, to refer to Derrida's expression,[19] is both the awareness of a common belonging to animal-being (which is only possible after recognizing an irreducible similarity in the heterospecific's diversity) and the animal-other I have become through the initiation rite that allowed my human identity to emerge. This third option—compared to von Uexküll's usable and Heidegger's objectivable-as-such—allows the non-human animal to go from being an extraneous entity "other-than-oneself" to being something "other-with-oneself" within the human area of reflection.[20]

1.2 From Epimethean Predication to the Promethean Meta-Predicate

The Epimethean predicate is primarily an important indication of a relation, and only secondarily it denotes a specific action.[21] In other words, a wing mirrors the dimension of flight both in its necessity aspect—what is needed to stay suspended in the air—and in its creativity aspect: the swallow therefore has margins of creativity within the dimension of flight, just as a human who is doing paragliding, and

[19] J. Derrida, *The Animal That Therefore I Am*.

[20] It is an "other-with-me" and not an "other-than-me". To understand the interspecific difference it is therefore essential to understand this proximity, therefore I will explain shortly what are the characters that bring us closer to the other species.

[21] von Uexküll already partly knew this, as his notion of *Umwelt* gives us the idea of a habitable dimension that is not only perceptible. J. von Uexküll, *Foray into the Worlds of Animals and Humans: With A Theory of Meaning*.

in these margins the swallow is the owner of its wings, endowed with the flight but not as a chain or imperative. The swallow is not captivated in flight because the flight itself provides for the creativity of the subject. The *Umwelt* does not amount to fixating the swallow against the sky but to freeing the swallow in the sky through flight.

To treat the Epimethean condition as determined is therefore a mistake justified by the idea that:

(a) the *Umwelt* denies creativity and that inhabiting finitude does not allow for singularity in the act;
(b) different *Umwelten* are separate monads with no possible overlappings or translations;
(c) the human does not have an *Umwelt* of its own, being totally free from declinative predications and living in a boundary-free world;
(d) functional predication does not allow one to distance oneself from being and always makes the latter emerge as usable.

Therefore it is clear that the mistake lies in reading the Epimethean not as an endowment grounding singular expressivity—the predicate as the construction of a dimension that allows for the animal subject to express itself—but as the opposite. Here we find the fracture between the human and the non-human animal: in the idea that the former is creative insofar as it is free from an *Umwelt* while the latter isn't. This line of reasoning is fallacious for two reasons:

1. if the human being is the outcome of a phylogenetic process then its constitution is necessarily predicative so that, like other species, it has an *Umwelt*;
2. it is not possible to express creativity if one has no predicate of expression, which is why the presence of an *Umwelt*—obviously understood differently from von Uexküll—is a necessary condition for creativity itself.

However, let's take one step back and reflect on Epimethean predication. If we evaluate the morpho-functional and expressive characters in humans and other animals, we must admit that the animal category as oppositional to the human being does not exist. The fallacy of interpretation must therefore be attributed not to the phenomenal analysis of heterospecificity, but to a later time, which introduced a gap between humans and other species. In the phenomenal analysis of being-animal we just recognize a collection of species:

1. that can be grouped in one category able to meta-predicatively define the needs-universalities of animal-being, such as heterotrophy;
2. that can be divided in more subcategories of a varying degree of proximity;
   (a) by taxonomy or phylogenetic closeness, where similarities (homologies) can be attributed to a common progeny
   (b) by ecology/lifestyle or functional closeness, where similarities (analogies) can be attributed to adaptive convergences.

The principle defined in animal-being therefore fully includes us and cannot be used as a dichotimic operator. In case (1), belonging can be defined by a common biological dimension, as animal organisms are characterized by particular bio-functional features (heterotrophy, sensiency, reactivity) shared by humans and all other zoological species, even those who are very distant from the phylogenetic viewpoint, such as invertebrates. On the other hand, in case (2) belonging must be expressed in a segregative way, i.e. for similarity of *bauplan*, namely in the form of predicative electivities that have different gradients according to the proximity of the two species considered with respect to the common ancestor (homologies) or in relation to the overlap of the selective morphopoietic pressures of the predicates themselves in the two species (analogies). In any case, both types of belonging cannot ground a categorial dichotomy between human and non-human animal because, following a homological logic, humans and chimpanzees (just to make the most obvious example) should be placed in the same category, as they share many more features than a chimpanzee may share with any other animal. The animal versus human opposition is therefore ungrounded as to the merely phenomenic or predicative aspect.

Thus, the Epimethean content represents the set of endowments of the animal-being to express its subjectivity, but it is not a dimension precluding the encounter with the non-human animal. On the contrary: it defines the encounter's electivity based on homological and analogical segregative coordinates. If we consider taxonomical proximity, we will realize that all mammals (including humans) have the same predicative project based—to make a few examples—on the infant's dependence on the mother, on parental behaviours used also as social and epimeletic expressions, on playfulness and metalinguistic communication in the ludic framework. Within this segregative dimension (i.e. within the project's main characters) it is possible for every mammal to exceed its own *Umwelt*: for example, a dog can recognize an epimeletic[22] need on part of a kitten, implementing a trans-specific adoption. On the other hand, if we focus on lifestyle we'll notice that all animals building collaborative communities (from humans to wolves) must develop behaviours related to help, mediation, distribution of tasks, operative consultations—just as it takes wings to fly, or fins to swim.[23]

The Epimethean sign is not to be interpreted as something species-specific isolating every species within its *Umwelt*, but rather as a signifier that, distant and difficult as it may be, can be translated thanks to its own Rosetta stone. When we compare ticks or bees to humans we do it by willingly increasing the distance between them, so as to enhance the humans. The universe of animal biodiversity allows the human being to always choose the right animal—passing it off as an exemplary—summarizing the antinomic character of animality and supporting the

[22]For "epimelesis" see glossary.

[23]It is precisely for this reason that a dog is able to enter into a collaborative relationship with a human.

pre-judice of distance.[24] By choosing in various animals—as allowed by the Epimethean predicative collection—the right predicate, we build an overall image of animality that is completely antinomic to that of humans (which is also reconstructed by appropriating non-human characters).

Every reference to anything in common with animals is used to ground prejudice, which is not only full of mistakes but also heavy with consequences. There is no such thing as an oppositional category of animality because it makes no sense to state that a chimpanzee has more in common with a jellyfish than with a human. And yet we fall into categorization every time we think there is a condition that characterizes humans beyond the peculiarity of the species—after all, every species is different by certain features—acting as a "meta-predicate" for which human beings are not included in the animal condition. Declaring either the existence or non-relevance of the human Epimethean character means precisely appealing to a meta-predicate, inaugurating a fourth kingdom—the human. In fact, if the predicate doesn't work one has to do the opposite thing and call for a lack of declinative predicates (an Epimethean deficit) in humans.

In fact, the meta-predicative structure does not define the animal peculiarity of the human species, but denies human belonging to the animal condition, kicking Epimetheus out of the human home. This is an artifice used as a grounding *passepartout* by humanism, in the attempt to separate the human by means of an unbridgeable hiatus. The meta-predicate can be configured in different ways—as happened in the history of thought—and therefore requires being unmasked every time. However, it basically always answers to an ideal of human transcendence. Such transcendence does not necessarily concern the soul or unearthly conditions, but indicates a predicative non-presence or non-contingency in the human compared to the animal—which, vice versa, is completely wrapped up in the here-and-now, that is, in the vital and bodily expression of its predicative declination. The overlap between the animal condition and Epimetheanism (i.e. the boundary to a function-range that rigidly establishes the expressive scope) turns the non-human into an entity exposed to the world, unable to be a protagonist in it.

Little matters that this profound presence of the animal in the moment/context is depicted (scornfully) in terms of captivation or (admiringly) immersion in the living. What matters is the discrimination itself—the separation of humans from animality, even if the latter is described as true, happy or what have you.[25] The animal is regarded as "closed in a space-time bubble"—the moment and the

---

[24]Thomas Nagel takes the bat as an example to declare the unknowability of the animal, while von Uexküll uses the tick to prove the perceptual-operating determinism of non-humans. T. Nagel, "What is it like to be a bat?", in *The Philosophical Review*, vol. 83, n. 4, 1973, pp. 435–450. But continuing with this logic, one could make many more examples. If we want to emphasize the violent character of animals we will use the best-known predator but we won't use herbivores. If, on the contrary, we want to emphasize the character of cowardice we will refer to the sheep and certainly not to the lion.

[25]These attributes, upon closer inspection, remind one of the aphorism "timeo danaos et dona ferentes" (Beware of Greeks bearing gifts).

world-context—completely open to the profoundness and the details of its present, incapable of abstraction or reference, and therefore passively expressing its drives and reacting to stimuli. The Epimethean being as described in the humanistic tradition is captivated in its own nature: its apparatuses are not endowments but imperatives, as the animal is unable to use them freely but has to follow the strings like a puppet.

The bubble enclosing the animal, according to humanist thought, is not only the perceptual-operational space (as von Uexküll recalls) but mainly a bubble of "here-and-now" from which there is no escape, as the animal cannot avoid the drive-reaction instinct. The animal's time is therefore a series of instants, not a flow, but an eternal present. Incapable of memory, of referring to something that isn't there, to project itself in the future, it cannot be bored due to non-going and it has no expectations. Condemned to the present, the animal has no biography (as it would require the sense of time continuity) and has no self, as its exposition to the world is total. Some don't even grant it emotions, not even fear or pain, as—they say—it gives itself completely and unconditionally.

The meta-predicate nullifies any predicative recognition. The discriminative tautology lying within humanist thought is not tied to a specific principle and can use different characters every time: biological incompleteness, eccentricity, language, self-consciousness, temporal disjunction or non-contemporaneity between signifier and signified, the awareness of death, or the notion of the truth—just to make a few examples. What matters is that the chosen condition can lead to the pre-humanistic distinction between a Promethean human being and an Epimethean animal, thus creating an ontological gap separating man from any other animal. For humanism, there are two different lineages. While Epimetheus connects the animal to the strict necessity of living through functional (connective) predicates, Prometheus frees humans from functionalities and from being entirely comprised in the context and the present.

Prometheism as a way to make the human emerge does not concern (as is usually believed) the human techno-poietic[26] tendency but rather its need for a correlative mediation: such is the function of tools in the absence of a specific declination. The Promethean character is usually divided in two main proposals:

1. *ab-origine* or phylogenetic Prometheism
2. *ex-post* or ontogenetic Prometheism with a meta-predicative function. The latter allows humans to be free from the "here-and-now".

Therefore, the Promethean meta-predicate allows humans (unlike any other being) to distance themselves from the importance and urgency of the moment, to refer to something non-present, to leave the instant, to exceed any time-space bubble, to say no to stimuli and drives. In this reading, animality is therefore Epimetheanism, i.e. full comprehension of the moment, singularity of vital expression, lack of reference to the non-present. Animality means a

[26]For "techno-poietic" see glossary.

meta-predicative condition that can be summed up in being totally captivated in the prison of one's own nature, so that categorisation does not refer to the simple "bearing of different characters" of the human. Whether we represent the Epimethean condition as vivid presence, vital poignancy, total opening to the world (as in Rilke's *Duino Elegies*)[27] or as captivation and poverty in world (ad in Heidegger),[28] it doesn't matter. Epimetheanism turns animality into a de-subjectified condition, because by breaking the continuum of life there ceases to be experience and life, and it is not possible to speak of biography, of, of expressive protagonism. In this reading, the human escapes the animal condition by being unrelated to the context, either as phylogenetically lacking a predicative conjugation—as underlined by Arnold Gehlen[29]—or as ontogenetically educated in language to distance itself from the here-and-now, referring to something absent. By leaving the moment, the human becomes absent, avoiding the calls of the world and the urgencies of the body. Prometheism creates a duality that humanism specifies in different ways through oppositional couples: mind-body, culture-nature, response-reaction, language-communication, true-functional.

When the symbolic—i.e. what lies between signifier and signified—emerges, the human can appeal to being no longer by means of indicative designating but by references: that is, to use Vygotsky's example, "to speak of the snake in the snakes' absence".[30] Distancing becomes the qualifying feature of human beings, the only ones able to escape von Uexküll's bubble and assume a universal perspective on the world. Non-presence allows for a sort of perceptual-operational *epoché*, a detachment from the world-context which turns into freedom of expression and autopoiesis. As a matter of fact, von Uexküll's monadology could not stand an accurate analysis, be it evolutionary or ethological.[31] Species-specific *Umwelten*—provided they exist as von Uexküll presents them—are not separate monads but overlapping

[27]"The creature gazes into openness with all/its eyes. But our eyes are/as if they were reversed, and surround it,/everywhere, like barriers against its free passage./We know what is outside us from the animal's face alone" Rainer Maria Rilke. *Duino Elegies*. translated by Robert Hunter with block prints by Maureen Hunter (Hulogosi Press, 1989).

[28]M. Heidegger, *The Fundamental Concepts of Metaphysics*, Indiana University Press, 2001.

[29]An evident example of this can be found in Arnold Gehlen's philosophical anthropology, which claims to interpret the human cultural dimension as exonerative compared to a lacking biological condition. In Gehlen's thought we find the humanistic leif motifs expressed in *De dignitaries hominis*, the sort of manifesto by Pico della Mirandola, where he imagines (appealing to a creationist vision) man as created outside of a biological rank, that is, devoid of performative declination and therefore protean and autopoietic. If Pico's image is understandable if contextualised in its time frame, it is no longer so after Darwin's discourse. (Hrsg.): Arnold-Gehlen-Gesamtausgabe, Bd. 4: Philosophische Anthropologie und Handlungslehre, Klostermann, Frankfurt am Main 1983.

[30]*The Collected Works of L. S. Vygotsky*, 1987, found at https://www.marxists.org/archive/vygotsky/collected-works.htm.

[31]Charles Darwin noted this already in his essay *The Expression of the Emotions in Man and Animals*, 1872.

entities.[32] But this is not the point. In the humanistic transfiguration, the concept of *Umwelt* becomes a way to support the disjunctive interpretation if man is taken to be the sole living being outside of the Epimethean bubble. To be outside the bubble means not having it, not being comprised within an *Umwelt*, inhabiting the *Welt* and being a world-constructor. This is how meta-predicative distance, for Heidegger, rejects any diversity, be it quantitative or qualitative: between man and other species there is an ontological abyss, signed by Prometheus.

1.3 Epimetheus' Forgetfulness

The perception of something missing in the human body is what grounds the myth of Prometheus and Epimetheus. Such narrative is understandable as an *ex-post* explanation: human beings, used to cultural shelters (a vast universe of tools, techniques, styles, domesticated biotypes and social protection), removed them and looked at their bare flesh, seeing themselves as naked, miserable, vulnerable—a mollusk without its shell. Looking around they saw self-sufficient animals, never naked, because dressed in a flesh fully declined to their species' performative needs. The non-human animal needs no addition, it is perfect in itself and everything more would be spurious and inauthentic. It already has the alpha and omega of its own existence, with no regrets or excesses, as all it is a fruit of nature. Humans see the animal's thermo-regulating fur, its fangs and claws to tear and wound enemies and prey, its wings to fly and its fins to swim, as well as its colours allowing it to emerge from or fade into the surroundings. With an improper operation, suddenly there is no longer the non-human animal with its predicates, but there is the animal world luxuriant with predicates.

Epimetheus' gift thus seems to overwhelm humans, because the comparison is no longer between the human and another species, but between the human devoid of its cultural apparatuses and the whole animal world, made of wings, fins, beaks, fangs, claws, feathers, fur—an enormous Moloch collecting all Epimethean characters as opposed to (and threatening) the human. Animals are performatively specialised—their bodies are tools perfectly apt for their lives and survival. At the same time, being specialised means having a fixed home in nature, a habitat. On the contrary, human beings lack specialised organs and are a real biological *monstrum* without a habitat. Therefore, if man wants to survive, he must create his own life conditions, changing the state of nature by means of artificial devices. Hence the concept of "compensation" expressed in Prometheus' gift, making up for his brother's forgetfulness. This is how the myth is presented and recovered by Herder,

[32]It is true that each species is immersed in its world-context. Such immersion is not only sensorial but based on motivations, emotions, knowledge, specific cognitive functions. We are mistaken when we say that different *Umwelten* are separated, because they overlap: there are wide areas of sharing between species.

who romantically resorts to the myth of the autopoietic[33] man realising himself in History. Thus, the Epimethean myth is characterised by cyclical a-historicity, whereas the Promethean world is marked by the linear path of history, to which man was destined.

The myth of Epimetheus and Prometheus, thus interpreted, holds very well and is even obvious or banal—it even resists the threat of Darwinism, albeit with an illegitimate interpretation.[34] The reason for this is clear: we are used to giving explanations *ex-post*, building narratives whose result is a "revelation" indicating a sort of predestination of hidden finality that should be brought out. This is the interpretative mechanism found in Kipling's *Just so Stories for Little Children*, which Stephen J. Gould describes as an epistemological obstacle.[35] Following such reading, lungs stand for fish's wish to leave water and feathers for the theropod's desire to fly. Thus, "just so stories" lead us to believe that man can invent a tool to compensate for his deficiencies, covering himself because he feels naked, building a house because he lacks a habitat.

The a posteriori explanation is simply wrong—it what in psychological terms is called teleological bias or finalistic heuristics. As we shall see in the next chapter, it is indispensable to revert the meaning of the sentence to understand the real causal direction: man feels like he is lacking because he got used to achieving things with tools, he feels naked because he is used to covering himself, he no longer has a habitat because he has built himself a house. It is the structure that allows for the performance, not the other way round. If today we tried to explain the invention of personal computers starting from the analysis of social networks we would never understand how this technology was born. It is the computer that allowed for the Internet to be invented, it wasn't the lack of something like the Internet that prompted people to invent the computer. Similarly, the ancestors of amphibians were able to leave the water because they had structures suitable to the function of gas exchange, which then were specialised into lungs, and birds were able to fly thanks to the feathers which developed in theropods for other bio-functional

[33]For "autopoietic" see glossary.

[34]I shall not dwell on demonstrating why the human species cannot be regarded as incomplete or lacking by presenting clear, evident, simple biological proofs—which, after all, can be inferred from the very mechanism of evolution. I wish to underline that I reject Darwin's revertive explanation recovered by Patrick Tort in *L'effet Darwin, Sélection naturelle et naissance de la civilisation,* Paris, Editions du Seuil, 2008. Here Tort claims that selective pressure is not the struggle of the strongest or the best fitted, but the slow process through which the real hybridizes a life project.

[35]G. Bachelard, *The Formation of the Scientific Mind: a Contribution to a Psychoanalysis of Objective Knowledge*, Cinamen press 2001. For Bachelard, the "epistemological mistake" is not a real mistake but rather a knowledge structure typical of scientific thought, in which the scientist deforms the reality she is analysing because of her interests. Subjective involvement thus biases and impedes the scientific spirit.

reasons.[36] The teleological explanation is nothing but a form of epistemic anthropomorphism: the attribution of a subjective will to a process.

The domestication of wolves can be a very good example. Today, looking at the way dogs behave with people, we might think that humans domesticated dogs to achieve exactly this type of behaviour, but this explanation does not hold for a number of reasons. The first and perhaps most obvious one is that most—in my opinion all—of the dog behaviours would have been of no use to a Palaeolithic human: conducting a flock wasn't useful because sheep hadn't been domesticated yet; the same goes for carry-over, as killing of wild game at a distance was not in practice yet; and obviously property guarding wasn't needed. But the even more compelling aspect is that the man could not have imagined, by observing the characteristics of the wolf, that it was possible to "derive" from it what today we call "dog"—also because the emergence of dogs is far from obvious. Finally, as reconstituted milk was unavailable in the Palaeolithic, and as wolves are not sociable, the only way to adopt a wolf is by materning[37] (having a woman milk a wolf), which is impossible without an Epimethean disposition. So the reason for the wolf'sadoption and domestication does not correspond to the dog's performative capacities. We can state that the dog's presence allowed for its uses and not the opposite. Once again the causal relation should be inverted.

However, this myth can also tell us something else, which is also interesting with regards to animal epiphany. The Promethean compensation undoubtedly reveals a sort of undisguised envy that the human being feels towards other animals. After all, every species has its points of excellence but also its shortcomings, and cannot think to be perfect, not even in its adaptive specializations. After all, what does "perfection" even mean in an evolutionary logic? Nothing. Evolutionism is the very negation of the concept of perfection, which only makes sense in an essentialist view of nature. When claiming that man is incomplete and lies outside the Darwinian paradigm, one is either adopting essentialism or being self-contradictory. Incomplete compared to what? What is the benchmark? In what sense could it be complete? The morphology of each species is nothing but an attempt in the great evolutionary flow, more or less endowed with fitness (a predicate that is never absolute or stable and never assessable in an e-nucleative way). It is a more or less static equilibrium in time, with no winners or losers but only changes.

Then what does this perceived perfection (of an eagle, a horse, a fish, a bee or any other animal) originate from? Why see an injustice at the origin of all things? Why come up with the sin of Epimetheus who, like an unfair or foolish parent, leaves a son without inheritance, filling the other riches? At a closer look, an eagle

---

[36]Think of the concept of "exaption" proposed by Stephan J. Gould and Elisabeth Vrba, "Exaption. A missing term in the Science of Form" in *Paleobiology*, vol. 8, no. 1 (Winter 1982), 4–15. The authors define exaption as "A character, previously shaped by natural selection for a particular function (an adaptation), is coopted for a new use—cooptation" (Table 1). An example is that of feathers, which are not only meant to allow for flying, but also for the subject's thermal balance. They are adaptations for thermoregulation and ex-adaptations for flying.

[37]For "materning" see the glossary.

has not that much to brag about for its nature, which often forces it to starvation and tremendous hardship, and a gazelle is far from lucky, living in the constant fear of being killed. Why then scorn the poor Epimetheus as if he were responsible for terrible injustice? The reason will appear very clear if we consider not so much the predicative gap between humans and other species, but rather the admiration and envy that humans feel for other species and especially the epiphanic meaning of the encounter with the non-human animal.

Speaking of animal epiphany means regarding the encounter with the heterospecific as pomading something beyond the merely phenomenic aspect: something involving profound and intimate aspects of the human being, a way to recognise oneself while denying oneself at the same time. Animal epiphany takes place when the human encountering the non-human animal feels an ambivalent sense of identification, i.e. projection and sympathy towards the animal as a shared being, and of admiration/repulsion for it, with vertigo and detachment. Animal epiphany questions the human being, it contaminates it as otherness invades the human's morphopoietic space. The epiphanic animal has already begun to affect anthropomorphy, showing other ontological dimension that are not just non-human animal as other-than-oneself but also viable paths—or better, paths that would be both fantastic and terrifying to walk. Animal epiphany is not the mere presence of the animal other but the phantasmic apparition of man become animal—the animal that has become part of you in presenting your possible projections (it doesn't matter if they are positive or dreadful). In this sense, the animal other has become other-with-oneself.

Animal epiphany makes the human feel the sublime and, as we shall see, any experience of the sublime can be traced back to animal epiphany: the shamanic trance of animal-becoming, the original sin in the Jewish tradition, the Dionysian chorus in Greek tragedy, hypsous in rhetorical style, the panic participation in nature, the mystical ecstasy of leaving one's body, the XVI century scientific counter-intuition, the technological infiltration at the end of the XX century. Animal epiphany is the manifestation of metamorphosis, the non-confirming mirror, Dorian Gray's picture—pushing us to reflect on ourselves and showing the body's nakedness. To see yourself naked you must first imagine to be dressed and this revelation is exactly what I call animal epiphany. Thus it is clear that there can be no envy without self-recognition in animal otherness and without being involved in heteromorphy. Only the human being that has discovered that it can be theri-anthropized, glimpsing at another possible dimension for itself, can envy the Epimethean predicate. Cursing Epimetheus is the same gesture (apparently reversed) of being expelled from Eden: it is the sudden break with one's own nature, a distancing from the body: it means seeing oneself naked and lacking something.

On the other hand, it is evident that only from a decentered position one can see oneself: we do not see the image in the mirror, but the image looks at us instead, because in the mirror we reflect on our body and not on our image. It is also clear that only thanks to a decentering otherness it is possible to reach a position from which to see oneself. In this sense animal epiphany means being absorbed in a heteromorphous dimension, almost detached from one's nature, seeing oneself naked, unfit, inadequate. Epimetheus' forgetfulness thus emerges at the same time as the human therianthropic projection, so that the Promethean transformation is not so much a compensation but rather the realisation of the epiphanic apparition—the emergence of the overhuman. And since Epimetheus is the one being stigmatised as responsible for the decentering process—the tree of knowledge able to reflect man beyond himself and colluded with non-human characters—then the anthropopoietic meaning of animal epiphany is evident. When the non-human animal becomes the other, the human discovers its humanity.

1.4 The Animal Mirror as Anti-narcissus

According to Ovid, Tiresias foresaw that Narcissus would have died old, had he never known himself. Therefore, Narcissus is condemned to never see himself if he wants to survive: he must have a direct, unreflective relationship with the world. But is a waterhole really enough to see oneself or even know oneself? If we borrow Pirandello's leitmotif, we'll notice that to be aware of one's identity one has to enact it in the expressive space given by otherness. Only through the others can the protagonist discover her characteristics, so that it is always possible to reinvent oneself by simply changing neighbours. And yet, paradoxically, since the expressive space of identity is defined by otherness, the reflected image is always spurious—an outcome of the elections that otherness operates on the subject's hundred thousand expressive possibilities. Thus every reflection is a dialogue and Narcissus' problem is not the waterhole but his dialogic incapability—not the fact that he sees himself but rather his inability to know himself. If this has any plausibility at all, it follows that every identity feeds on otherness, be it an individual, cultural or generically human identity.

Traditionally the mirror is considered such if it doesn't deform the image, representing it faithfully and not altering the identity. That's why the expression that takes place in a dialogic milieu is taken to be a stretch of the identity, the inevitable contamination leading to inauthenticity. In this reading, reflection is understood as an objectifying operation whose aim is the recognition of identifying characteristics in the search of the purity of the self. But can we really speak of an authentic self if the very term implies a reflective (i.e. contaminating) process? The question could be rephrased as follows: are we reflected in the other because we are used to doing it in reflective surfaces—be they mirrors or waterholes—or is it rather the other way round? In our ontogenetic experience our first mirror was undoubtedly our mother, and in order to be reflected in her we had to project ourselves in her, seeing her as a *Doppelgänger* able to establish centrifugal coordinates in the world as well as areas of proximal growth.

The reflective experience cannot be ascribed to a neutral surface: it is rather the acquisition of a contaminating space. If we apply this to the emergence of identity mentioned above, we'll notice that reflection—as an unveiling of the self which is hidden by non-expressiveness—implies a series of requirements which are far from obvious. For example:

(a) the presence of an otherness able to provide the self with an expressive space, which is inevitably elective;
(b) the subject's de-centred position, which is a pre-requisite of looking at oneself since, as long as one is Narcissistically immersed, there can be no reflection;
(c) an inter-subjective encounter able to activate projection and detachment, the first step of the emergence of identity as a gap between the I and the self.

If we consider these three aspects we won't fail to acknowledge that the principle of self-reflection is not a self-centred and autarchic process but rather a dialogic event. Every form of mirroring in the traditional sense of the term (i.e. seeing one's image reflected in a reflecting surface) is actually a subtype and a consequence of the habit of being reflected in otherness.

Reflection does not start off as an introspective event but as a sympathetic process, under the aegis of a "feeling together",[38] in a dimension where there is no real gap between the self and the other, who mutually project and compare each other. After all, the child constructs a self-image through a process of co-sensience with the mother in which sympathy plays a fundamental role. The mother is a second *Umwelt* for the child, capable of bringing out certain beings over others and define inner operational models. It is no hazard to speak of maternal epiphany, which is able to make the child's self emerge by means of doubling. Thus the child becomes confident about de-centring, participating in the others' emotional and motivational states. In the projective-participative dialogue with the parent, the child builds her identity. Thus to reflect means to see oneself through the other. In the emergence of her individual identity, the child comes to have some self-awareness thanks to her relationship with her mother, who is the first reflective being she comes across.

Thus, the reflective entity is a neighbour we can project ourselves into and compare ourselves to. Identity needs otherness in order to define itself: so, for the construction of individual identity you need an individual otherness, for a cultural identity you need a cultural otherness and for an anthropological identity you need a non-human otherness. Reflecting on the human position means seeing oneself through the non-human animal and building an interface of transpacific projection. In this sense man could and can reflect on his anthropological identity only by the virtue of a non-human otherness experienced as a projective continuum. That's why the epiphanic encounter with the non-human animal makes reflection possible and, therefore, allows for human identity to define itself. If this is the case, then it is evident that to understand human identity one needs to define the type of reflecting entity—i.e. the animal mirror: one must establish its features, its way of reflection, the kind of deformation it produces, the transpacific interface.

[38]The American philosopher Ralph Acampora calls this "symphysics". R. Acampora, *Corporal Compassion: Animal Ethics and Philosophy*, University of Pittsburgh Press, 2006.

On the other hand, if it is true that without a mirror you cannot see yourself, it is also true that in reflection you project yourself beyond yourself. Speaking of reflection as de-centring or as a flow of de-centring processes (a series of splittings) helps us understand the importance of a projective bridge between us and the reflective entity. Identity emerges when I can see myself and leave the total co-presence and instantaneity of my dimension. In this process we must inevitably admit the importance of de-centring to reflection: it is not the mirror that determines the splitting, but it is the projection into the mirror that allows for reflection. In fact, I do not look at the image, it is the image that looks at me. Only a human that is de-centred in the animal mirror, that is, in the condition of animal epiphany, can observe herself from a human perspective.[39]

On the other hand, the emergence of identity produces new levels of belonging which require to be externally fed. Climbing up the mount of identity creates a heteronomic dependence, facilitating heteronomy processes. Seeing oneself in a therianthropic key determines an ontogenetic space to which the subject is projected. Therefore the reflected image should be considered a place of tension rather than a stable and definite position: using Vygotsky's words we might define it a "zone of therianthropic growth."[40] In the same way, the maternal reflection represents an evolutionary space: tension and not a finite product. The over human image produced by therianthropic reflection is therefore an evolutionary objective that presents itself coherently with the species-specific dimensional learning but that, unlike the latter, introduces elements of differentiation within the species, creating processes of segregation. When I speak of a project implicit to the reflection I do not at all mean an explicit objective in the human being's hands, but a proximal area of growth that is not regulated by species canons.

The therianthropic reflection refers us to the image of a hybrid human being, projected towards the over human. Usually this condition is regarded as a salient character of the human being and rightly so, as it creates a difference between species-specific contents and human contents. Therefore some anthropologists differentiate ontogenesis from anthropogenesis as the construction of the human dimension that cannot be traced back to phylogenetic contents: all in all, even though I'm a bit sceptical about regarding humans as special in this process, I agree. The point that stays unanswered in traditional anthropology is how to find a reason for such process while prescinding from God. If anthropogenesis produces a way to go beyond the phylogenetic dimension, it is not possible to hypothesise evolutionary autarchy for human identity—it would mean hypothesising that the anthropological Narcissus is able to see himself without the waterhole. If the emergence of a human identity—like all identities—involves a reflecting dialogue with otherness, then there is an absolute need for an antidote to the human's

[39]As I said, in this sense the myths of Adam and Prometheus are absolutely identical, if we interpret the act of *hybris* characterising both as the image of the therianthropic emerging from the epiphany.

[40]L. Vygotsky identifies the relationship with the mother as the emergence of a proximal level of experience: a supporting structure for the subject's overall development.

Narcissistic narcosis as an ontopoietic dimension. There is no doubt that anthropopoiesis involves something different from the social-maternal learning typical of species endowed with parental care and not deferrable to the experiential occurrence typical of declarative events grounding individual identity-variability. If anthropopoiesis has to do with the human condition, it has to be a dimensional event able to construct the human dimension, therefore diverging (if not emancipatory) from the species canon. So? Who can be the maker of this process?

The authors underlining the half-breed nature of cultures and the vacuity of the claim of a solid and pure identity, the ones that recall the importance of exchanges with cultural otherness and affirm the dialogical importance of the biographical structure of the individual—these are the ones who stop in horror at the trans-specific Rubicon, refusing to apply to the human dimension the same operator of reflection with otherness they see in the construction of every individual and every culture. There is no doubt that anthropopoiesis creates a de-centring from the species canon, introducing an anthropo-decentring ontological perspective.

So who has the role of identity interlocutor, if not the otherness/counterpart of man—i.e. other species? They are able to determine projective and distancing effects and are both recognisable and different. The hypothesis of decentralisation—i.e. of an exit from the phylogenetic gravitation—only holds if we admit that there is an external intervention that can't be traced back to the species predicates, able to outline a new area of proximal growth based on an epiphany showing what is possible, showing the future. That's why an anthropology that does not ground its explicative foundations on transcendence (going from a theocentric to an anthropocentric anthropopoiesis) has no de-centring power: that is, it has no way to subsume the anthropopoietic process. This is the role of animal epiphany.

Chapter 2
The Promethean Condition

To rediscover the non-human animal as a plural geography (which requires being aware of the many hermeneutical frameworks surrounding the concept of animal) it is therefore necessary to reflect on the human. To rediscover the non-human animal we must free it from the antinomic concept of animal, but to do so we must extract the disjunctive operator from the bundle of fractalic recursivities. I will dwell on the humanistic paradigm, considering it not so much a XV century line of thought but rather a philosophical mindset excluding nature in general and non-human animals in particular. In my reading, humanism was defined in philosophy and post-sophistry, then grew between the I century B.C. and the I century A.C. through stoicism and monotheism, later feeding on neo-Platonism and gnosticism and finally being definitively affirmed in Italian humanism.

Humanism is characterised by a strong anthropocentric tension requiring that one:

(a) removes all that concerns the human—both in its ontological definition and in philosophical inquiry—any reference, contamination, overlap, contiguity or metamoprhy with non-human animal otherness;
(b) removes any therianthropic element from any original mythopoietic[1] structure, from the image and expression of creator divinities and from the eschatological reflection.

All that concerns the human, both in the reconstruction of the origins and in the projection of a meaningful future—both on earth and in the afterlife—assumes an anthropomorphic image. On the other hand, precisely in the attempt to distinguish the human from the non-human animal, humanism tends to reduce to the idealised image of the anthropomorphic everything that might appear to be shared with other species. The outcome is that humanistic anthropocentrism does not place man (understood as a phenomenic unit) at the centre, but rather an *idea* or vision of man, which is paradoxically opposed to man in his natural characters. Thus organic features, motivational dispositions, the needs and pleasures of the body, perfor-

[1]For "mythopoietic" see glossary.

R. Marchesini, *Over the Human*, Numanities - Arts and Humanities in Progress 4, DOI 10.1007/978-3-319-62581-2_2

mativity unmediated by a tool, the direct contact with nature—all these things are marginalized, hidden, ignored, neglected, if not completely denied. This happens by means of a hierarchization of predicates placing at the top everything that highlights the difference between humans and other species.

Humanistic anthropocentrism does not correspond to anthropocentrism proper as an anthropocentred dimension, because while the former focuses on the emancipatory idea of *anthropos* as opposed to *Homo sapiens*, the latter is based on founding man on man. Anthropocentrism proper aims at absolutizing man as such in a centripetal way, so that the human gravitates on the constitutive characters of *Homo sapiens*. Therefore, the two forms of anthropocentrism proceed in opposite directions. However, humanism manages to pull an amazing trick: it makes Epimethean shortcoming into the gyroscope capable of maintaining the orientation axis of emancipatory centrifugation despite the mutability of anthropocentric rotation. If the lack of predicates is the phenomenic realisation of man as the project of nature, then there is no incoherence between the emancipatory position (humanistic anthropocentrism) and the praiseful one (anthropocentrism proper). The lack of predicates is used to sustain the idea of human freedom and self-determination as well as to support the hypothesis of autopoiesis[2] and non-determinism. The Epimethean deficit, determining a lack of a specific placement in nature, is assumed and hypostasised by humanism *à la* Pico della Mirandola to turn man into:

1. a neutral, non-declined entity able to act as the measure of the world—hence anthropo-metrism[3]
2. a virtual identity able to assume any performativity thanks to technopoiesis, therefore destined to be the container of the world—hence anthropo-plastics.[4]

If *De hominis dignitate* can be regarded as the manifesto of humanism, Leonardo da Vinci's Vitruvian Man is probably its best iconographic expression. The idea that the human being is an entity devoid of adaptive endowments (a kind of abortion of nature because devoid of those predicates that lead each species to excel within their own ranks) turns man into a virtual entity capable of assuming any form without being performatively declined. According to Pico, the human being oscillates in its magmatic and protean condition, able to reach the dimension of angels as well as to fall into the telluric realm.

What is proper of man is attributable to this keeping the distance from the world, maintaining a pure condition allowing him to reach the hyperuranium he is destined to. In this reading there is the obvious rejection of any form of pre-determination as the founding principle of an autopoietic conception of the human being. At the same time there is an evident need to construct a anthropo-poietic project through the comparison with the other species, so that humanism takes the oppositional category of the animal condition as its basis. This means that if the Promethean

[2]For "autopoiesis" see glossary.

[3]For "anthropo-metrism" see glossary.

[4]For "anthropo-plastics" see glossary.

meta-predicate falls, then the whole humanistic building will crumble with it. That's why (contrary to Charles Darwin's continuative evolutionism, von Uexküll's *Umwelten* monadology, or Konrad Lorenz's ethology linking expressiveness to phylogenesis) the humanistic system always brings forward new meta-predicative characters aimed to keep the categorial dichotomy between man and animal alive. Animality is therefore a humanistic contraction with very variable boundaries, so that it can adapt to the human's needs to define its own identity.

2.1 A Second Nature

The *principium individuationis* of the human as found in Greek culture oscillates between:

(a) the concept of logos, which unites rationality and language, leading to seeing the human being as "sapiens"—that is, different from the animal due to its ability of self-reflection;
(b) the technopoiesis, i.e. the propensity to performative externalisation with the aim of outlining a being endowed with technical intelligence, who is a *faber*.

On the other hand, this very oscillation between self-reflection—Rilke's reversed eyes—and the operative distance made possible by tools in turn generates a further interpretative dualism, one that sees the human as:

(c) lacking a rank and therefore exposed to the world and to pleasure—hence the Dionysian tendency;
(d) virtual or tending to stay in an embryonic, sketched condition, thus potentially autopoietic—hence the characterisation as "creator" untied from any directive, free and fully in charge of its own destiny.

These four main points are adopted by humanism, which makes of ontological anthropocentrism its foundation. The animal is therefore relegated to the role of background against which the human emerges loosing the predicative dimension that was still present in Aristotle's definition (*zoon logon echon*) and definitively acquiring the meta-predicative function. The Epimethean predicate is regarded as a synonym of opacity, stiffness, limitation, restriction, rootedness. Therefore, affirming Prometheism means:

1. considering man in the light of "the human" as an elevated condition compared to the animal;
2. regarding human nature as transparent, plastic, virtual, released, eradicated. The image that humanism proposes oscillates between Botticelli's embryonality and Vitruvian projection seeing the human as ascending and autopoietic.[5]

[5]On the one hand, man tends to elevate and emancipate himself from the world using the animal as a representative of the latter, on the other there is a phagocytosis annihilating or removing meaning from everything that is not human, transforming it into a more or less usable object.

As we have seen, the humanistic revolution turns the human being into a Vitruvian *vir*, who enters the world and shapes it, dictating its metrics and morphology. Making the world anthropomorphous means interpreting it according to one's own measure units while re-projecting it according to one's own proportions. The non-human animal as a form, proportion, or measurement is thus declassed aesthetically, epistemologically and ethically. Humanism magnifies man as a morphopoietic principle: *kalokagathia,* that is, order, cleanliness, lightness, transparency, brightness, roundness, smoothness. On the contrary, the therianthropic represents the opposite and has to be expunged. In this sense Pico della Mirandola's proposal places pre-humanistic mysticism in the body as the expressive and representative dimension of man. Pico's nine hundred theses in his *Oratio de hominis dignitate* (1486) are an exhortation, in addition to stating the superiority of humans over animals. The human is described by Pico as a virtuality that requires an effort of evolution through philosophical (that is, autopoietic) practice. It is interesting to recall the passage in which Pico della Mirandola explicitly defines the direct relationship between God and man and that between man and other animals.

> We have given you, O Adam, no visage proper to yourself, nor endowment properly your own, in order that whatever place, whatever form, whatever gifts you may, with premeditation, select, these same you may have and possess through your own judgement and decision. The nature of all other creatures is defined and restricted within laws which We have laid down; you, by contrast, impeded by no such restrictions, may, by your own free will, to whose custody We have assigned you, trace for yourself the lineaments of your own nature [...]. We have made you a creature neither of heaven nor of earth, neither mortal nor immortal, in order that you may, as the free and proud shaper of your own being, fashion yourself in the form you may prefer. It will be in your power to descend to the lower, brutish forms of life; you will be able, through your own decision, to rise again to the superior orders whose life is divine.[6]

As you see, human freedom basically lies in not being an animal. Another important scholar related to the humanistic proposal is Marsilio Ficino. For him, man is *copula mundi*: that is, the intermediate and connective term between the divine and the earthly, the place where the two meet. Man realises the mediation between earthly life and transcendence, thanks to a privileged position that grants him freedom. In Ficino we especially find the themes of Hermeticism and gnostic reverie appealing to a principle of revelation leading to philosophical and religious syncretism. In XV century humanism, human dignity and human superiority over other animals are at one, insisting on the oppositional dichotomy for which the non-human is determined, positioned, non-free and incapable of understanding and creation. Man's freedom from theocentric oppression requires an ontological scapegoat, which is the non-human animal crushed into the condition of brute—something to avoid and keep away from. The brute is completely immersed in the telluric that does not allow for any levitation and strictly binds it to a niche, while the human being aspires to free itself completely, becoming virtual so as to reach

[6]G. Pico della Mirandola, *Oration on the Dignity of Man*, Henry Regnery Company, Chicago 1956, p. 7.

another dimension. This is possible thanks to a different visualisation of its own form, which has to become diaphanous, light, and protean.

It should come as no surprise that Descartes sees animals as automata, as it is the humanistic view that triggers the antinomy between man and animal. Underlining the animals' complete immersion into nature, their incapacity to distance themselves from the "here-and-now", their bond to the contextual present, all of this inevitably produces a mechanical conception of the non-human. If it's true that the Copernican revolution, defeating geocentrism, opens the door to infinite spaces making the human a small part of the universe, it is also true that this very fact made it even more urgent to find a new gravitational centre for the human. The XVII century scientific revolution can therefore be interpreted in the light of this oscillation: on the one hand it questions the anthropomorphous reading of the world, on the other it confirms the role of man as it was depicted in the humanist manifesto (that is, the domination of the world through reason and *techne*). This confirms the idea of mathematising the non-human universe, which underlines the gap between the human being (as a free entity capable of inhabiting history) and the rest of nature as a whole (as an isochronic entity following linear and deterministic causes). Newton definitively marks this dualism: the human dimension progresses in the historical development, while the non-human is subtracted to the progress of time.

The Promethean spirit is confirmed by the frantic XVII century technical evolution which led humans to think they could explain mechanically the whole universe, feeling like demiurges of a second nature *in fieri*. Think of hydraulic automata adorning German gardens, clocks marking time and serving as a dualistic metaphor between being in time and having time, wind instruments and organs able to reproduce the voices of animals, the mills turning praxis into production: all these things confirm the idea that the animal is nothing more than a machine, more complex that human technology but still a machine. Their externality becomes therefore their casing, hiding their internal mechanisms—a metaphor of sensory deception as proof of the wound inflicted by the Copernican revolution. Prometheism thus means leaving the sense interface and finding shelter in an inner monologue, only believing in the echo of one's own thought, abandoning the last fragments of contiguity with the world.

On the other hand, the world is shaken by new rhythms and new horizons under the blows of the techno-scientific transformation. Promethean becoming is the realisation of history: a second nature laid open before modern man (a man who is in his time and has made his time his own). Prometheism is more and more synonymous with becoming, transformation, acceleration, history as a new salvation. Fire, which used to make the world malleable by conforming it to the *anthropos*, is not dynamicity, promised light, future. Thomas Moore's utopia, blending the prefixes *eu* and *ou* and therefore not necessarily projectable to another dimension, becomes exclusively outopic, turning the whole a-historical reality into a dystopia.[7] After all, ever since the XVII century, external references have appeared less and less stable and adventure in new Meccas becomes a common topic, a custom. History is thus the stage on which Prometheus is able to perform new distancing goals for the human lineage.

The stress on history as a human condition as well as knowledge is underlined by Giambattista Vico, who, while refuting the solipsistic principle of the Cartesian *cogito*, places history and human creativity in a relationship of mutual foundation. Making history and dwelling in history undoubtedly refer to Promethean recursivity, where *techne* assumes the role of flywheel in this self-catalysing process of disjunction. Outside the historical dimension, man appears both innocent and indecent, because he belongs to a total immersion in a progressive flow. The Enlightenment is also unable to distance itself from this reading, underlining man's childish dimension and aiming for the education to critical rationality. Prometheus takes the role of educator: he brings out the universal hidden inside the contingent. Prometheus is the enlightening reason, challenging the obscurantist divinity and its evil emissary Epimetheus is the capacity for abstraction. When reason is asleep it creates monsters, as in Francisco Goya's famous painting, and monsters are always therianthropic.

On the other hand, if infinity gives us vertigo it is also because (and especially because) it puts in check the Promethean light trying to illuminate it, then the need to focus on man becomes even more urgent. Besides, it is precisely in the XVIII century that the conditions arise to strengthen the anthropocentric vision that will find its greatest and deepest endorser in Kant. In his 1785 *Groundwork for the Metaphysics of Morals*, Kant emphasises the difference between a morality we could call "local", referring to the diversity of the people and contexts of life, and morality proper, based on rationally founded and therefore universal principles. In 1798 with the essay "Anthropology from a pragmatic point of view",[8] Kant refines this concept by underlining the importance of focusing not on man as a physiological entity but on what man, being free, can/must do with himself. Even if Kantian cosmopolitanism sets him against Romanticism, the introspective and isolationist trend is the same, if not stronger, as the one emphasising humanistic anthropocentrism.

The two great industrial Revolutions of the XIX century (the first can be traced back to the steam engine, the second to electricity) support the idea of man's

---

[7]The problem with the humanistic paradigm lies in its anthropocentric project based on exalting man as the sole protagonist endowed with existential activity, as opposed to the inert and passive non-human. The humanist paradigm is grounded on a substitution that, respecting the canon of Medieval theocentrism, simply puts man in God's place. In this perspective, the human being can have an outopia: an ontopoietic goal placed elsewhere thus devaluing nature, regarded as dystopic with respect to human predicates. Thomas Moore's utopia, which blends *eu* and *ou*, is not necessarily placed elsewhere. But in being assimilated into humanist thought, it loses the suffix *eu* and becomes synonymous with Neverland (which demands from man an inevitably anthropocentric formalization). The outopia (henceforth "utopia") thus becomes a negation of nature or an urgency to get rid of nature.

[8]I. Kant, *Anthropology from a pragmatic point of view*, Cambridge University Press, 2006.

enclosure into a second nature created by Prometheism. Machines start proliferating, they gradually become new pets, in an evolution destined to surround the human universe like a sphere. The technosphere becomes paradoxically more human than man himself, because it adheres more to Prometheism, so that at the end of the XX century man dreams of turning into a machine. The machine judges the flesh, it establishes the standards to be met, inverting the old ergonomic paradigm. But in the XIX century these trends were still latent, even though this Golem appears as Frankenstein in Mary Shelley's book. The machine is what opposes the animal, overcoming the topos of St. George and the dragon—the machine is what wins over the animal by pushing it into its telluric abyss. Through the machine, the human being is taken to another world characterised by becoming and progressive projection, finally realising the scenario hinted at in Pico della Mirandola's manifesto. History gradually becomes ascensional.

For Johann Gottfried Herder, the analogical-correlative dichotomy between man and animal is based on the fact that, while the animal world is nature, the human world is history: a sort of autopoietic second nature. In his "Treatise on the Origin of Language",[9] Herder states that every human idea is not an immediate product of nature, but rather represents man's own product. In this sense, the human being detaches itself from nature, producing his own works himself. Herder has the same conception of distance as Pico, so that man is realised in history, where he finds his habitat, his real dimension where art and language represent the most authentic expressions of the human, revealing the individual and the people. For Herder, the ability to distance oneself lies in the predicate of reflection understood as gathering and freedom to dwell on an image. Prometheism and narcissism—as self-referential mirroring, unlike animal epiphany which is a hetero-referred mirroring—converge.

Herder's celebrative view of man is made explicit in his *Outlines of a philosophy of the history of man*,[10] where the philosopher restates the idea that the human being is not only the measure of reality, but also its goal, referring to it as the "flower of creation". Once again man must stand out from the green land of animality. Herder is the first to clearly notice man's biological shortcomings by relating them to Prometheism, as a compensation but also an opening to the world. This finds new strength in idealism, thanks to the importance given to the subject compared to phenomenic reality. Man is increasingly closed in his amnios, up to annihilating what lies outside of consciousness. The Kantian I-think is not just the transcendental incipit of knowledge, it is also the source of being, in terms of both form and content. The human being progressively identifies himself with Descartes' *res cogitans*, operating an isolation that nullifies any contiguity with the non-human and establishes the dualism (already latent in Descartes and then re-proposed by a series of dichotomies like nature-culture, instinct-reason, cyclic-progressive, facts-values) embodying the never-ending conflict between the Epimethean world

[9]J.H. Herder, "Treatise on the Origin of Language" in Herder. *Philosophical Writings*, ed., Michael N. Forster, publ. CUP.

[10]J.H. Herder. *Outlines of a philosophy of the history of man*, London 1800.

(centred on nature, instinct, cyclicality, facts) and the Promethean world (characterised by culture, reason, progress, values). Prometheus promises the human being completeness, a utopian horizon for which to sacrifice the animality left in the human: its flesh. And that's why, in the great ideologies of the end of the century, man is tied to the mountain and Prometheus is the one eating his liver.

The XIX century also witnessed great changes in the evaluation of the history of life. Once again, the relationship between man and animal lied at the centre of the discussion. If Linnaeus had placed human beings together with other primates, it is thanks to Lamarck that the bases were set for a continuist theory of the species, perfected in Charles Darwin's *The Origin of Species* (1859) and *The Descent of Man*, as well as *Selection in Relation to Sex* (1871). Darwinism truly is embarrassing for humanism because it seems to question the distinct genealogy proposed in the myth of Prometheus and Epimetheus. Consequently, soon people started to push the English heretic back into the humanistic frame:

1. one can reject Epimetheanism by considering it vestigial and turning contiguity into a risk of regression, as exemplified by Mr. Hyde and many other therianthropic characters (a narrative topos of the *fin de siècle*), or into an indicator of ancestrally (as in Lombroso's physiognomies);
2. one can amplify the gap between man and animal by rejecting the very principles of humanism, rejecting the latest areas of overlap, as does Heidegger in the *Letter on Humanism*;
3. one can reinterpret or assign a particular meaning to Darwin's evolutionism, for instance in a teleological anthropocentric sense (so that human beings are the final goal of the process); or one could regard the evolution of some human behaviours as lying outside of evolution, as Patrick Tort does.[11]

To turn the common ancestry between humans and animals into an irrelevant acknowledgment, it is necessary for humanism to push it back to remote past up to hiding it completely. And yet the Darwinian wound still hovers like a ghost responsible for building a bridge between the human and the non-human, so that the latter, albeit theoretically, can now come back to life and contaminate the former. This is why the therianthropic is unable to re-establish the principle of non-human diversity and becomes a regressive polarity: the animal is a dark mirror, an ancient baboon that can re-emerge (mister Hyde or Conrad's Kurtz) or the beastly ancestral (Zola's Jacques Lantier or typology in Lombroso's physiognomy) and inaugurate new forms of racism.

The animal as a regressive symbol turns the non-human animal into something belonging to the remote past, a non-present presence, a living testimony of the (horrible or nostalgic) past. It nullifies, or claims to nullify, any contiguity, reference or dialogue between the human and the non-human. The animal thus doesn't manifest itself through the exhibition of a *quid*, but it is rather what man no longer

[11]P. Tort, *L'effet Darwin, Sélection naturelle et naissance de la civilisation* (Editions du Seuil, Paris, 2012).

is—the animal is the fall. Once again, the humanistic paradigm refuses to think of the non-human animal, sticking to the antinomic categorisation of the animal as a background.

Existentialism goes in the same direction by denying the non-human's effective existence. Heidegger sees the concept of *Umwelt* in the animal world as a state of total absorption in the contextual here-and-now—hence the captivation characterising the non-human as opposed to the distancing and therefore the subjective expression qualifying the human being. Heidegger's reading of the concepts of *Umwelt* and openness are different from their proposals in von Uexküll and Rilke. As per the *Umwelt*, the bubble is the state of ontological conjugation of living beings (including the human being) even though there is a hierarchy or an emergence allowing the human to exceed it. However, this proposal becomes philosophically contradictory the very moment man claims to define the other *Umwelten*. Heidegger solves this inconsistency by freeing the human from the *Umwelt*. Like a player who realised he's about to lose—in this case, Heidegger realises that the humanistic predication is no longer enough to make man special—the German philosopher raises up with a sort of ontological bluff, denying any animality to the human and widening the gap with the non-human animal.

Heidegger proposes an overcoming of humanism through an anthropocentric acceleration that increases the distance between the human and the animal. After all, the *Umwelt* reminds one very much of the idea of the animal's Epimethean closure and can easily be assimilated to the humanist paradigm. Thus Prometheism becomes the dimension of living in the world, or being a "constructor of worlds"—an idea in which the autopoietic and performative principle of *techne* is evident. Heidegger's open signifies not being completely absorbed in the world's direct usability, distancing oneself from the object—which is the only way to see it as an object-in-itself.[12] Man differentiates himself from the non-human animal, because he is not animal, and every relation to the non-human animal, is lost because what appears is only the animal phenomenon—which, albeit in its biological

[12]Martin Heidegger writes in *The Issue Concerning Technology and Other Essays*, Garland Publishing, New York and London 1977, p. 4: "We ask the question concerning technology when we ask what it is. Everyone knows the two statements that answer our question. One says: Technology is a means to an end. The other says: Technology is a human activity. The two definitions of technology belong together. […] The manufacture and utilization of equipment, tools, and machines, the manufactured and used things themselves, and the needs and ends that they serve, all belong to what technology is. The whole complex of these contrivances is technology. Technology itself is a contrivance, or, in Latin, an *instrumentum*. The current conception of technology, according to which it is a means and a human activity, can therefore be called the instrumental and anthropological definition of technology." This is an instrumental view of technology as a mere activity reifying any human production. However, Heidegger seems to imply that it also represents a dialogue between the outside world (to which being is open as being-in-the-world) and the subject itself, in a performative and binding relation. In this perspective, *techne* appears to be able to craft the human, modifying and hybridising its body, predicates and perceptivity of the outside world. The non-human partner, be it technological or animal, makes the human referable, in constant need of external support to define its predicates and its being-in-the-world.

multiformity, can be neither an interlocutor nor a mirror, as there is an ontological gap between us and him.

The third response to the Darwinian continuum is to differentiate the morphopoietic process of non-human animals from those that led to the species *Homo sapiens*. In this case it is believed that, while animals are the outcome of the continuous, inevitable and refined work of selection—understood as the law of the strongest—human beings are different. The evolution of social behaviours, which emerged through selection, has managed to save man from natural selection itself. There is no doubt that Darwinism has led to very problematic social applications.[13] After all, today it is clear that these proposals were the outcome of a mistaken interpretation of Darwinism. Considering human predicates to be the outcome of selection, with the latter being interpreted as the law of the strongest, leads to believing that every action against such law will lead to a progressive worsening of the predicates.

However, some of the answers to these aberrations, based on the same hermeneutic specularity, lead to results that are just as wrong, by supporting the idea of man as predicatively indefinite. In other words, one takes shelter even more in the humanity paradigm by:

1. confirming the somatic-expressive non-specialisation of man implicit in the myth of Prometheus and in Pico's manifesto, taking it as the human *quid*;
2. believing that such lack of declinations is the result of our species not being subject to natural selection.

This inversion maintains, or claims to maintain, Darwinism within humanism by stating that the human being owes its qualities to the very fact of being sheltered from natural selection. In this case, Prometheus is interpreted as the tutelary deity that protected humans from the selection and Prometheism becomes the ability to break free from the selection mechanisms that would lead to a deadly specialisation.[14] It is acknowledged that humans are particularly inadequate when they are born—still, compared to other carnivore puppies, the latter are much more immature than humans. Another consideration concerns small size of the human follicle mantle or the splanchnocranium but, as we shall see, it is much more plausible to think of these effects as the outcomes of a slip of selection than as masking and neoteny.[15]

As early as 1864 Alfred Russel Wallace underlined that, after the development of pro-social and intellectual faculties, physical superiority was no longer awarded

[13]Think of Herbert Spencer's social Darwinism or Francis Galton's eugenics.

[14]A hypothesis that goes in this direction is that made by Lodewijk Bolk, who believes that the human predicative primitivism can be attributed to neoteny, that is, a delay in the development process for which the foetal conditions persist until the adult age. Another similar proposal can be found in the concept of "proterogenesis" by Otto Schindewolf and in that of "domestication" by Konrad Lorenz. For "neoteny" see glossary.

[15]Patrick Tort precisely and accurately defined this view of masking in the concept of revertive effect operated by pre-social behaviours.

because the weak and the sick receive assistance. As is clear, even for Wallace it comes to defining a meta-predication for the human being, definable by the term "evolution of evolution". Darwin also stresses the importance of the development of the social instincts that, associated to memory and remorse, allow human beings to pursue a different evolutionary path. As Patrick Tort notes, Darwinian anthropology does not manifest any break from evolutionary mechanisms, but rather a "braking effect". In other words, one might posit the existence of an exception to performative selection—a suspension of the mechanisms of adaptive specialisation—by the virtue of the evolution of pro-social behaviours. Patrick Tort's proposal has several weaknesses, but I will only underline three:

(a) the idea that selection is a universal power like gravity, acting on all organisms as a struggle of all against all;
(b) the idea that there can be a time without selection, excluding a species from the parameter of fitness;
(c) the idea that the selective process should produce results concerning the organism's ability to eliminate the others and win against competitors.

These legacies do not regard Darwinian thought, but rather the interpretation of Darwinism based on existing cultural frames. First of all it makes no sense to speak of natural selection as a singular entity, reducible to the Hobbesian image of the affirmation of the strongest: fitness, i.e. the capacity for differential reproduction, can only be referred to a single species. One must refer to a variety of selective pressures, where each shielding—but it would be more correct to speak of "slippage"—does not cancel selective processuality, because the reproductive differential is always in force, but simply moves it on other predicates. In this sense phylogenesis is always the production of predicates. Finally, it is incorrect to think of evolution supported by selective pressures as the emergence of egoism, struggle, or oppression: in fact, in certain biofunctional contexts, fitness—the true parameter of the production of predicates—can be linked to opposite behaviours. One example is the evolution of parental care, where success is achieved by devoting oneself to the other.

2.2 The Sharp Separation Between Human and Non-human Animals in Philosophical Anthropology

In principle, however, up to the 1940s, Darwinism was widely rejected in favour of the idea of a totally different genealogy between humans and other animals. An example of this is the philosophical anthropology by Max Scheler, Helmuth Plessner and Arnold Gehlen, which is also related to some philosophical currents such as hermeneutics, phenomenology and existentialism. This line of thought plays a particular role, a strategic one, with regards to the matter under discussion, precisely because it is centred on a decidedly Epimethean view of the animal—which is enclosed in von Uexküll's *Umwelt*—as opposed to a lacking human being relying on Prometheus' kindness. As you can see, philosophical anthropology is much in debt with the humanistic manifesto, to the point that it is sometimes defined as "neo-humanism", as well as with Herder's Romantic theory.

The attempt to relaunch the humanistic idea of man as a special project while avoiding the manifestly creationist aspect of Pico della Mirandola's thought take place through a complex interdisciplinary process aimed at conciliating human science and philosophical reflection. Given the importance of authors like Scheler, Plessner and especially Gehlen to the topic under discussion, I believe it is indispensable to dwell—albeit not exhaustively—on these authors and, in general, on philosophical anthropology, which is still very popular today.

Philosophical anthropology invents nothing compared to the foregoing considerations on the relationship between animal Epimetheanism and human Prometheism. However, it is also clear that, dealing specifically with the human sciences and reconciling philosophical reflection with the latter, philosophical anthropology should be considered the true term of comparison in order to refute:

1. the autopoietic and self-reflexive (narcissistic) conception of man (and animal epiphany aims precisely at this, considering man as a hybrid and not as self-founded;
2. the discontinuist view of the phylogenesis of the species *Homo sapiens*.

The difference between the human being and the animal is very significant in Max Scheler, who talks about a spiritual principle, complementary to the vital one—shared with animals—that makes man a unique and exceptional entity, radically different (again in a meta-predicative way) from all other living things. For Scheler the human being is characterized by living in a sort of connective agape, where the *ordo amoris* becomes in fact the *principium individuationis*. The element that connects the person to the world is therefore a state of openness that is revealed in values, i.e. proto-phenomena that direct the behaviour of human beings. Scheler makes explicit reference to the concept of "bildung" as a formative tension of the subject in culture, which is in effect the ecosystem that characterises the human being.

In this sense, also the tool in Scheler is not just a tool, because its ultimate meaning is attributable to its being a "way to culture", one that allows the human being to position itself in the animal-bodily totality in the dimension that is proper to it. In his 1928 essay *The Human Place in the Cosmos* Scheler claims that while the animals "always say 'yes' to reality (…) the human being is the 'Nay-sayer', he is an *ascetic of life*."[16] In Scheler there is an evident tensional aspect—the same we found in Pico della Mirandola and Marsilio Ficino. Man is realised by escaping a purely functional or adaptive dimensionality, so even what at first sight could be explained in view of a merely functional end—the tool—becomes a kind of

[16]M. Scheler, *The Human Place in the Cosmos*, Northwestern University Press, 2009, p. 39.

instrument of fitness that enables man to practice in the cultural dimension. The ascetic view of man as the nay-sayer—living within an agapic dimension, self-formed and exempt from Hobbes' *homo homini lupus*—needs the animal as a counterpart that, vice versa, has to follow its drives, deploying its individualistic and selfish aggressiveness in the struggle for life and being unable to act on itself as pre-constituted by the laws of nature.

Prometheism apparently takes the form of compensation when in fact it represents the banner of self-determination and the release from any outside contributions: the green light to a dimension of unconditional power of self-projecting. That's how the animal, represented as something external to stay away from, actually becomes a disjunctive operator called to erode humans from the human project. The disjunctive meta-predicate is identified by Helmuth Plessner in the concept of "eccentricity"[17]: while animals live at the center of their environment—bounded by von Uexküll's bubble—and therefore their lives are governed by a strict determinism, the human being is in ex-centered, by virtue of the self-awareness that allows it to deny, and move away from, the biological necessity which, on the contrary, the animal cannot escape.

In this sense the animal is seen as devoid of a self or of a self-awareness and, because of this, it is assumed that it cannot leave its bubble, thereby transcending the biological gravity of its nature. Therefore, for Plessner the human is the realm of freedom and choice, whereas the animal is entirely bound to necessity (it cannot do otherwise) and the determinism implied by reactivity and drivenness. The meta-predicate of eccentricity places the animal condition in *being* a body, while the human one coincides with *owning* a body—that is, being able to use it and think of it. For Plessner the human dimension is also a sort of original condition: a kind of primary detachment from organic life from which the gap between man and animal derives.

According to Plessner, the human being is inserted into his own bodily limits but, at the same time, is projected outside of them, opening up to the world and therefore acquiring an "ex-centred" position. However, precisely from this eccentric position the human being traces new poetic coordinates, that is, new artificial boundaries to embody. This being-a-body and being-in-a-body inevitably changes the human condition compared to the animal one, as in the psychic dimension projected outside (but it's unclear where) man is inevitably decentred from his body, so that he considers it one of the many things in the world. Human conscience is therefore interpreted as non-coincidence with the body: a laceration of the body, a fracture with respect to the bodily dimension. At the same time, human schizophrenia produces a reflection that amounts to self-reflection.

In Plessner, the break of unity is conscience but also disquiet, creative unbalance, process. There are undoubtedly some points in common between Plessner's conception of the human ex-centred position and my proposal of animal epiphany:

---

[17]Plessner, Helmuth. 1928. *Die Stufen des Organischen und der Mensch: Einleitung in die philosophische Anthropologie*. Berlin: Walter de Gruyter & Co.

they share the idea that human evolution can happen under the aegis of anthropodecentralization. However, there are still obvious differences lying in:

(a) the idea that ex-centration produces reflection, whereas in the concept of epiphany the reflection it is the therianthropic that produces the decentering process;
(b) the autarchic idea of a process necessarily based on the leitmotif of the human scarce Epimethean endowment, which leads to conceive of reflection as self-reflection, whereas in animal epiphany only the encounter with otherness can realise a non-narcissistic reflection, and therefore an escape into the overhuman.

As you can see, both in Scheler and in Plessner, the *ab-origine* human diversity compared from other species is never questioned, and the humanistic notion of human autonomy in its self-construction is preserved. If we consider Pico's manifesto, we'll notice that even God is kept out of the anthropopoietic process, because His operation merely sets a starting condition for the human being, allowing for its own self-determination. The basic principle of humanist thought is ontological anthropocentrism, which does not contemplate any non-human interference (be it a divinity, the technosphere[18] or animal otherness) in the construction of the human dimension.

Humanism can never do without narcissism (i.e. self-reflection), but any theory that tried to explain decentralisation starting from an isolationist ontopoietic conception is obviously bound to be weak. Therefore, even in Plessner reason falls within the usual meta-predicative dualism used to distinguish mind versus body and human versus animal. In 1928, the same year as Scheler's essay, Plessner's *Die Stufen des Organischen und der Mensch* also came out. In the book, Plessner compared the vegetal, animal and human dimensions based on their inner cohesion and autonomy, noticing a progressive graduality in their ability to distinguish themselves (i.e. their ability of self-decentralization). While the vegetal is completely absorbed in its position, the animal is capable of a certain autonomy that sets it apart from, and in opposition to, the environment. But while reacting to environmental stimuli, it is not aware of what it does because, unlike the human being, it remains anchored to its own center, thus not being able to "see itself from the outside."

However, Arnold Gehlen is surely the thinker that brings the metapredicative conception to its most complete formulation. For Gehlen, the human scarce predicative endowment deprives man of the screening that the animal has thanks to its full inclusion in its *Umwelt*. Thus, man is literally overwhelmed by stimuli, while the animal only follows those relevant to its life process. Also, the human being undergoes a flood of drives, and not only those that blindly and automatically lead

[18]For "technosphere" see glossary.

to adaptive behaviours.[19] Human indefiniteness or incompleteness make us projected towards perceptual and pulsional exposition as well as to ontopoietic plasticity. In fact, the human being can distance himself from the world thanks to language, conscience, action and technology.

This is the concept of exoneration, which has common roots with Plessner's eccentricity and explicitly recalls Promethean language, which like the centaur is appointed to head towards what is far away—in time and in space—unlike Epimethean animals living the uniqueness of the here-and-now. Thus for Gehlen the human being lacks Epimethean predicates and is therefore characterised by adaptive shortcomings, ancestrality or primitivism, non-specialisation, so that it cannot live directly in nature. Unlike animals, man lives a second nature. Experiencing the world, where the excess becomes distance, basically means that, while the animal has a specialization living in a particular environment, man has access to the entire world, but he must necessarily change it through his work. Through cultural action, for Gehlen, the human being compensates for its organic weakness, creating instruments that become extensions of its body, but he also builds a second nature or dimension that is no longer qualified only by performative analysis. Technology, which is central to Gehlen's thought, becomes the compensation for human deficiencies and non-adaptations, as an integrating part of man's nature.

If it is true that the need to mark the boundaries between man and animal can be found in the whole of philosophy, mostly after Pico's manifesto, it is also undeniable that the philosophical anthropology proposed by Scheler, Plessner and Gehlen represents an important break, which should be considered as going hand in hand with philosophy on the one hand and descriptive sciences on the other. What characterises it is the need to grasp the overall human "specialness", joining scientific research and philosophical reflection, considering the human being as a synthetic being that should be identified in its entirety. Many authors underline the neo-humanist meaning of this project, understood as a need to:

(a) synthesize the outcomes of the so called human sciences in the light of philosophical interpretation—hence the proximity to Gadamer's hermeneutics;
(b) overcome the trauma of Darwinism, guilty of finding a despicable continuity between man and other species.

For philosophical anthropology, as underlined by Habermas[20] and Coreth,[21] the human being cannot be reduced to a particular science, that is, no science can grasp

[19]In Gehlen's hypothesis, the original human lack makes man a foetal being devoid of protection and thus exposed to a number of expressive vanishing lights, which the philosopher interprets as ambivalent qualities. In fact, on the one hand they allow for the virtual action that underlies human creativity, on the other hand they represent social and adaptive dangers if they're not kept under control—hence the idea of Prometheus.

[20]J. Habermas, *Theory and Practice*, Beacon Press 1975.

[21]E. Coreth, *Was ist der Mensch?: Grundzüge einer philosophischen Anthropologie*. Innsbruck, Wien, München: Tyrolia 1973.

the essence of man. Thus we need a philosophical interpretation of scientific data hiding the intention to redefine the boundaries between the human and the animal. The idea is to give back the human being its global image, one that cannot be reduced to the empirical data of single sciences or degraded to a merely phylogenetic aspect. More coherently we might say that, without the Cartesian ground taking the *res cogitans* to be the human essence and with the affirmation of Darwinism, the humanist paradigm was suddenly unstable (as noticed especially by Martin Heidegger). Hence the need for a solution. For philosophical anthropology, we must place man at the centre, starting from his meta-predicative characterisation—its being special in a global way, as a particular project of nature—ignoring the humiliation suffered because of the Copernican, Darwinian and Freudian revolutions.

Thus, while animals allegedly differentiate themselves through Epimethean characters (so that a jellyfish is different from a chimpanzee only in terms of predicative complexity, with the latter living in a richer bubble) the human being belongs to another condition defined by the meta-predicate of absence or non-declination. The Promethean dimension should be read, both in philosophical anthropology and in existentialism, as the universal *quid proprium* of the human being as a lack of boundaries and as emergence from the invisible bubble enclosing the animal. The adaptive perimeter defines the animal's captivation, its complete abandonment to the environment and hence its perceiving-operating only in the adaptive sense, contrary to the farsightedness of those who can have an absolute gaze on the world, looking at it from its peak, far away from the particular, both contextual and temporal.

2.3 Is the Human Condition Original or Produced?

While philosophical anthropology considers the human condition of absence/distance as an original prerogative—the outcome of a project of nature—so that human self-creation (anthropopoiesis) is understood as a compensation for the poor Epimethean endowments, in Heidegger's existentialism human becoming (anthropogenesis) is inherent in the separation itself, or failure-fault, in the biological functional flow inaugurated by the representation of being. Anthropogenesis is a process that makes mankind able to separate the subject from the object through language. Here, however, language does not complete the human being nor does it compensate for her shortcomings: man himself is called to create the distance and make the object emerge in the subject's perspective.

For Heidegger language is what allows man to grasp being as such, separating perception from action, accessing the beings in the world for what they are and not for their biological-functional value. Distance separates the being from its function: if the bee sees a flower as nectar, man (only man) can grasp the flower as such. The mediation of the word transforms the relationship with the object in a non-inclusive interaction: if the bee is absorbed by the flower as a transition to the biological function, so that we cannot even speak of a real relationship between subject and object, the human being is capable of grasping the flower as a flower and not as a chance of finding nectar precisely because he distances himself from it. Distancing implies the recognition of the object as non-engaging: thus it is an object with respect to which one can show detachment or lack of interest, a neutral evaluation, as it is not "the implicit terminal of an action with a biological value."[22]

Heidegger's position undoubtedly can be ascribed neither to the first position (*humanitas* as the original Promethean condition) nor to the second, that is, the idea of *animalitas* as onto-genetically overcome or hidden by the symbolic distancing achieved by language. However, Heidegger's view is interesting because it deprives the non-human animal of a real perspective on the world, i.e. the ability to face the world as an observer who is not merely a user. In an almost opposite way, Derrida makes the cat emerge as a subject by feeling observed by the cat—when his nakedness becomes such because there is someone able to see it as such. The observing animal is exactly what is completely excluded by Heidegger's interpretation of "poor-in-world."[23] Thus Prometheus becomes the one who watches, the one who can observe because he has distanced himself from the world. Heidegger locates in language the principle of separation from the environment that marks both the human subjective existence and death as something beyond the end of biological functions. Language is therefore a mediating function between subjectivity and death—Derrida agrees on this.[24]

At this point it might be interesting to better analyse the difference between:

1. an *ab-origine* meta-predicate (as in the philosophical anthropology of Scheler, Plessner and Gehlen) considering the human (incomplete, lacking, embryonic) condition as a background to culture, i.e., the symbolic, technical, linguistic act;
2. an *ex-post* meta-predicate (as in authors like Lacan, Cassirer, Dennett and others) for which culture is what inaugurated the human condition, domesticating man's *animalitas* through the symbolic-linguistic system.

The difference, however, lies not in the metapredicative principle—*ab-origine* versus *ex-post*—but rather in the very meaning of *humanitas* operated by culture, which in the former case focuses on the process of a second adaptive space, while in the latter case focuses on the realization of the distancing. Plessner's position is maybe the most intermediate of the proposed ones because, while accepting the lacking condition that nature allegedly imposed on humans, he introduces the fracture of being in the body and also outside the body as a principle of reflection.

To understand the Promethean element, after all, one must start from the devaluation of the non-human and the nullification of the non-human animal. In the

[22]F. Cimatti, *Filosofia dell'animalità*, Laterza, Roma-Bari, 2013, p. 34.

[23]M. Heidegger, *The Fundamental Concepts of Metaphysics*. Here we also read "Not even the lark sees the open", a metonymical expression indicating that non-humans cannot see beyond their *Umwelt*.

[24]J. Derrida, *Of Grammatology*, Baltimore and London: Johns Hopkins University Press, 1997.

ab-origine conception, animality does not regard the human being because it was forgotten by Epimetheus to begin with. This is obviously a healthy forgetfulness for the humanist interpretation of humanity, because it allows for the freedom of action, self-construction and self-reflection that it couldn't have had otherwise. Therefore, for philosophical anthropology Prometheism is both freedom and uncertainty, power and insecurity, and yet Prometheus' legacy represents man's irreducibility to the animal. Animality is therefore a devalued condition, that is, minor, hierarchically inferior, notwithstanding the description of man as lacking. Animality is a counter-term to the protean form represented by Prometheus; it amounts to being closed in one's condition, while the human being is a person because it is plural: it is body, in the body, outside the body, able to reflect on every position it takes. If, on the other hand, animality is taken as a lost or hidden condition, then the devaluation lies in placing it in the past (an ancestral or infantile regression) taken away from the human by the symbolic, by the domestication operated by language and the irreversible fall into referentiality.

Starting from parental care, the human being is immersed in complex networks of semiotic mediation taking place through language and ranging from nominative emergence to prescriptive coordinates proposed by adults and introjected into ontogenesis. The very fact of assimilating social rules, definable in prescription resonances able to give directives to the body, understood as immediate expression of physiological and instinctual reactions (but does it still make sense to interpret behaviour this way?), transforms the human being into a schizoid entity because it creates a gap between the self and the body.[25] In this regard, quoting Cimatti, "transcendence appears in human life as the effect of language on the body, taking the subject away from the body, because saying 'I' means leaving the body and having the chance to see oneself from outside one's body—a chance that the tick seems not to have."[26]

Beyond the differences between the two proposals, there is a convergence in the conception of *humanitas* as a Promethean principle: some believe that the original Epimethean shortage has revealed the Promethean need, others consider the Epimethean emptying as a result of the Promethean emergence. The result is the same: what characterises the animal is Epimethean fullness, what characterises the human being is the Promethean second nature. The devaluation of the heterospecific, however, lies not only in the contralateral conception of the non-human but also in a blunt denial of any intervenience or even influence of non-human otherness in the construction of the human. The autopoietic arrogance of the human must rely on a real explanatory "stunt" to try to explain a decentring process. But how is it possible to decentralize oneself while remaining firm within one's ontological dimension? And how can one recognize oneself "from behind"—

---

[25]Lacan makes human subjectivity emerge precisely from this gap in which mirroring is self-reflection of the human, as it is completely bound to the human itself. Thus the Promethean ends up defining he who looks at himself in order to emerge.

[26]F. Cimatti, *Filosofia dell'animalità*, p. 40.

that is, in the sense of no longer what I was—if not through the emergence of a metamorphosis? Self-reflection inevitably closes the subject in what it is, namely in what remains. Only a non-narcissistic reflection, i.e. one that is not self-referred, is able to produce a decentring split and thus open up the human.[27]

Before addressing in detail the issue of animal epiphany as the overcoming of autopoietic narcissism, it is useful to consider some weaknesses of the two above mentioned proposals, both based on the exclusivity of the Promethean foundation. The attempt of philosophical anthropology to explain the metapredicative origin blatantly clashes with evolutionary theory, so it is not possible to find accordance between the two, not even by using imaginative epicycles such as that of the shielding on natural selection caused by the evolution of social behaviour. As mentioned, philosophical anthropology transforms anthropopoiesis into a kind of human dimensioning—the advent of an autopoietic second nature inhabited by man as unfit to relate directly to the first nature—transforming culture in a kind of completion or useful crutch to make up for performative deficiencies. If the concept of the deficiency of human nature collapses, the backbone of the whole theoretical edifice ceases to exist.

The argument that such supposed deficiencies are yet to be proven is being progressively abandoned in the growing awareness of the performative complexity and specialization of the human: after all, what does a central nervous system consisting of 100 billion neurons amount to, if not an adaptive specialization? However, the weak point of Gehlen's proposal is the idea of a side project, that is to say, one that does not follow the phylogenetic coordinates proper of other animal species. To think that a totally unfit species lacking functional correlations might evolve all of a sudden, as you can see, requires moving away from the foundational principle of the bios based precisely on adaptation: that is, on the correlation of the species to a certain function. For Darwin's theory, it is the species that builds its rank and not the other way round, which would lead to a fixist vision of the living, so Gehlen's fixist legacy is evident.

But then we must wonder whether some form of inclusion between the theory of incompleteness and Darwinism is possible, albeit with some adjustments to Gehlen's hypothesis. A useful starting point comes from Charles Darwin himself. In the essay *The Descent of Man*[28] he aims to analyze the relationship between the development of prosocial behavior in humans—evolved in accordance with the principle of fitness—and the effects of these compared to further selective processes. But even in this case there would be too many inconsistencies with the theory of incompleteness; for that reason, Arnold Gehlen's rejection of the Darwinian theory is inevitable and consistent, not a hiccup. In fact, it would not

---

[27]It then becomes clear that the most important aspect of Prometheism is the concept of human autarky I have already explained in *Post-human. Verso nuovi modelli di esistenza*, Bollati Boringhieri, Torino, 2002 and in *Il tramonto dell'uomo. La prospettiva postumanista*, Dedalo, Bari, 2009.

[28]Ch. Darwin, *The Descent of Man*.

make sense to speak of original human incompleteness being in an evolutionary logic, for several reasons. Let's see some:

(a) we cannot assume that primates, the most complex animals as per ethographc articulation, would lead within a few millions of years to the most incomplete and deficient evolutionary line in the animal kingdom;
(b) imagining the evolution and affirmation of a completely deficient being cannot be justified in the explanatory mechanism put forward by the Darwinian theory, based on the concept of fitness;
(c) if we observe the morphology and structure of some human biological characteristics—such as the eye-hand coordination, the complexity of the cortex, the biomechanical structure enabling bipedalism—the high specialization of the human being appears undeniable.

On the other hand, as we shall see, it is much more likely to feel that the original deficiency proposed by Gehlen is actually a "perception of deficiency": that is, a sense of lack or need attributable to the habit of performance mediated by technology, which sets standards of optimality that are unattainable without a tool. Similarly it is not enough to appeal to language as an entity that suddenly intervenes to realize the fault between the carnal body and symbolic body, or, more generally, between being and the body; one must try to understand—as Chomsky or Vygotsky do—why this process emerges in humans. However, the explanation given by these authors does not help support the metapredicative discontinuity they would like to argue in favour of. Language can increase the designation and therefore the distance from the entity, but it cannot inaugurate it from scratch. On the other hand, the metapredicative conception founded on human becoming-absent-separated-divided through the domestication of the human body (i.e. considering the anthropogenetic process as a distancing objectifying event that develops in the human experience when language comes into play) requires more attention.

As Vygotsky stresses, language is viewed as a means of distancing because it allows to refer to an entity in its absence. Language is itself capable of operating a split in the ontogenetic process and consequently making the object thinkable outside of contingency. The designation becomes a split because on one hand there is the signifier—the autonomy of the representation-symbolization from the entity —and, on the other hand, there is the entity itself, which for the first time may appear "as such" (that is, not as an extension-expression of the life drive of the animal). The world becomes a workshop for Prometheus, the educator that, through language, makes man the master of his own behaviour.[29] The entity as such maintains its presence both in the absence of the subject that it encounters and in the absence of a process of fruition. Language, in other words, would interrupt the flow of fruition that confuses the animal in the world, annihilating any form of subjectivity in it. The animal lives, it does not reflect on the entity: the bee does not relate

[29]As we have seen, for Heidegger language is the house of being, the place that allows the environment to become world.

to the flower but to its own vital function of sucking nectar. According to this interpretation, while fruition is the status quo characterising the animal world, the reflexive and representational approach characterises the human.

Prometheus is the principle that allows human beings to be absent, to create a fracture between themselves and the world which saves them from the stunning of intoxication. Prometheus is a faraway look, while Epimetheus is stigmatized as a form of myopia that involves and overwhelms. The animal, as Epimethean, is a de-subjectivized entity who lives the continuum of the expression of the body, and whose presence (non-present and especially lacking a present) always corresponds to the vital function that the body activates in the here-and-now as urgency and as conjugation. Animality provides a bubble of perceptibles and, within it, a set of markings that indicate the different predicates of the usable, which results in a perception that is already action. Twentieth century neohumanism created a rift between the animal unable to recognize the object, and the subjectivity of the human being all based on competence in focusing things across the distance.

For the animal, the entity is not an observable but a functional predicate—graspable, suckable, edible and so on—and thus it can never appear in its objectivity. The meta-predicate is therefore an ontogenetic result: the emergence of the human thanks to the distancing operated by language, finally able to focus on the body as such. What's at stake is the emergence of the human as irreducible to animality and irreversible: it is the appearance of a subjectivity that looks at the world and at its own body by breaking away from it. The consequences of the emergence of language are relevant both to the definition of subjectivity and to the judgment of the entity as such.

Regarding subjectivity, authors like Agamben, Chomsky, Vygotsky, and Dennett, for whom the metapredicative difference between the human—as a fatal condition of those who are no longer included in the condition of *animalitas*—and the animal is based on language, tend to see the symbolic designation of being as:

(a) the emergence of the self-that-speaks as an entity that owns the body, namely that is no longer a body, or that is capable of reflecting on its vital functions;
(b) the emergence of the self-that-observes, which sees the body as a whole and not only as the usability and adherence to the living entity;
(c) the emergence of the self-that-governs its own animality in relation with the outside world and bodiliness, that is, the self who is able to say 'no' to the stimuli coming from the markings of the world and to the impulses arising from its body;
(d) the emergence of the narcissistic-self, which is separated from the world. It looks at itself and takes awareness of being an "I", establishing itself as a biographical and subjective singularity, that is, as an "autonomous perspective"—as knowledge and relevance of a self that is positioned in proximity to, and outside of, the latter;

(e) the emergence of the self-that-dies, because only after dissociation can the subject be present and therefore die, while the animal positioned not in the present but in the singularity of the here-and-now can only stop living;
(f) the emergence of the self-in-distress not only due to the awareness of death but also to the very distance of living, the lack of a position or, if you will, of an answer giving reason for its existence.

Again, what interests us is the alleged discontinuity between the human being and the animal that, stripped of any title and subjectivity, is not able to provide a form of otherness for the human being. This reasoning is flawed by some weak points; however, it seems more plausible than the *ab-origine* metapredicative conception proposed by Gehlen. In particular:

(a) the idea that self-awareness and one's own biographical self, that is, individuality persisting over time, is linked to language and not simply perfected by it, since, especially in mammals, there is indubitable evidence of self-reference and self-recognition;
(b) the idea that the non-human animal is not able to categorise, that is, that the processes of categorization and conceptualization depend on the presence of language, even though the use of general categories of reference to beings has been proven not only in mammals but also in other types of animals;
(c) the idea that only language allows for self-control or the ability to cope with frustration and also the possibility of self-imposed limitations, even though it has been shown that parental care is a form of apprenticeship for social animals related to the implementation of processes of self-inhibition, such as bite inhibition in canines;
(d) the idea that the animal is completely absorbed in the singularity of the here-and-now, even though many tests on birds and mammals show that the individual has goals and therefore looks to the future, planning its action based on strategies/tactics, making simulations and assessments based on the structural requirements of the problem, and reflecting on finding useful solutions (heuristics) choosing them on the basis of their consistency with the problem;
(e) the idea that the animal is unable to assess beings as such: that it is unable to objectify reality, but only to use it driven by immediate urges, even though there is proof of the animals' ability to stock up objects even regardless of their immediate need for them.

But the most controversial aspect in assuming language as an anthropogenetic principle is that it makes the emergence of language itself inexplicable, as in any self-grounding claim one is always condemned to a *regressio ad infinitum*. If, however, we treat language as one of the many cultural expressions of the human—capable of educating the human being, but not an autopoietic and emanative product *of* the human—we can set language within the decentrative experience inaugurated by animal epiphany. Only when language emerges after the epiphany—for example in a zoo-mimetic act—as distancing from the merely species-specific expression, it introduces an ontopoietic dimensionality, for example the use of the phoneme for phatic purposes, making man increasingly decentralized within the designation. The close relationship between the mime, the gesture, the prosody, and the camouflage—detectable both in anthropological field research[30] and in the analysis of neurobiological structures—actually make it plausible to posit a relationship of co-optative development between practices that differ considerably in their modes of expression, but all related to zoomimesis,[31] and language development.[32] Again, as we shall see, we must reverse the explanation: language was not born as a designative and symbolic tool, but its emergence has made its functional emergence available.

[30]Consider the research by Irenäus Eibl-Eibesfeldt, clearly showing that the evolutionary legacy of *Homo sapiens* is far from deficient. In fact, our species is phylogenetically tied to a group of animals (anthropomorphous primates) that are some of the most complex ones in terms of ethographic structure. See Irenäus Eibl-Eibesfeldt, Human Ethology, Aldine Transaction, 2007.

[31]For "zoomimesis" see glossary.

[32]This is the hypothesis underlying Merlin Donald's. *Origins of the Modern Mind: Three Stages in the Evolution of Culture and Cognition* (Harvard, Harvard University Press, 1991). The idea is that the human mind evolved differently from the primates through adaptations that have led to a new representative system. Instead of replacing the previous modes, the latter has sedimented them, giving rise to a mosaic made up of different structures. For Donald the first transition happened with the passage from an "episodic" culture to a "mimetic" culture typical of *Homo erectus* and characterised by shared knowledge, rituality, the reproduction of events in writing, etc. The second passage would be from such "mimetic" culture to a "mythical" one present in all humans and identified by the use of verbal language and a highly developed semiotic skills. The last sedimentation allegedly took place with the advent of "theoretical" culture characterised by graphic invention, the formulation of theories and so forth.

Chapter 3
The Therianthropic Being as Our Neighbour

The encounter with the non-human animal is based on the act of seeing and being seen, since the animal is a "being-to-be-seen".[1] Alfred Russel Wallace marvels at the morphological beauty of the birds of paradise and their courting dance. He is especially astonished by the fact that these performances take place in dark and shady woods, "with no intelligent eye to gaze on their loveliness."[2] Anthropocentrically, Wallace cannot help feeling shocked in front of such a waste of beauty, thus expressing an echo of the humanistic principle, very popular in the Seventeenth century, for which the world is a huge theatre built for mankind. However, in the exhibition of its own form and in its specific rules, therianthropy finds its functional principle, which always also takes visibility into account.

The hidden and internal parts of the body do not follow the principles of symmetry, proportion, and colour combination that, on the contrary, are typical of all external organs.[3] This is no coincidence, since therianthropy is to be looked at. Undoubtedly the encounter between man and animal otherness becomes a way to confront, and conform to, each other. Therianthropy as an encounter of forms in a complex game of references, identifications, projections and centrifugations, is what interests me when I speak of otherness. Otherness, therefore, is not a foreign entity, not even when it is very far from phylogenesis and it's present in my experience only as a sponge or a sea star, sudden and new like Wallace's birds of paradise, or when it is unavoidably absent like trilobite or anomalocaris fossils.

I can see myself even in a skeleton or a footstep, since there's a shared principle, a sort of "synmorphy", even in manifest heteromorphy. Wallace knows the intrinsic beauty of the birds of paradise because he recognizes them as if he had already seen

[1]A. Portman, *Animal Forms and* Patterns*: A Study of the Appearance of Animals*, Schocken Books 1967.
[2]Wallace as quoted in *The Birds of Paradise: Alfred Russel Wallace: a Life* by William Bryant, iUniverse, 2006, p. 139.
[3]Portmann calls them "visual structures." See Portmann, *Animal Forms and Patterns: A Study of the Appearance of Animals*.

R. Marchesini, *Over the Human*, Numanities - Arts and Humanities in Progress 4, DOI 10.1007/978-3-319-62581-2_3

them, as if they were present among his parameters—that is, he recognizes himself in them. When Edward Wilson speaks of "biophilia"[4] he refers to the aesthesic pleasure in recognizing the animal form and the desire to project this form or, in other words, to transform aisthesis into aesthetics.[5] Animal-being consists in self-recognition and aesthetic sharing that go beyond the individual's specific perceptive universe. In fact, I use vision to speak of the condition preceding the encounter and enabling it only to ensure an easier comprehension.

Synmortphy[6] is an aesthesic thread that testifies to some basic parameters of animal beings. The theriomorphic being is my neighbour, even when I barely know it. First of all, my affinity to animals is the outcome of the awareness that we share a common ground—the animal that therefore I am, using Derrida's words—namely, a shared type of life: one that is proactive in the unbalance, fragility and vulnerability of existence. In animal otherness I find myself, I feel at home. And I know it recognizes me, interacts with me and expects me to react. This is not the case with any other interaction we have with the world: only animals recognize each other as neighbours. Addressing other animals is a feature of every animal: think of a hyena threatening a lioness to seize a carcass or a wolf trying to catch a deer. This reciprocal self-recognition is not a form of anthropomorphism.

A form of co-belonging is present even before the encounter, as there is something unsaid and implicit in the recognition of the game and in the wonder, fear, desire of animal otherness. Considering the non-human animal a *res extensa* is an intellectualistic operation or maybe a fantastic projection—a refusal of reality. Humanism is therefore an illusion or a hallucination; it may be pleasant and productive, as all dreams are, but you cannot dream forever, as every prolonged dream becomes a nightmare. I fear that the Twentieth century is the product of the Humanist dream turned into a nightmare.

The Heideggerian reading of the animal *Umwelt*, fully centred on what can be directly used and captivation, removes me from the animal gaze; it turns me into an empty shape, and creates an irreducible dissimilarity between us. Besides, this reading deprives the non-human world of all that cannot be used and makes it extremely fragile. In this way, perception is transformed into reality within a radical structure that simply absorbs the flow of what satisfies its needs. On the other hand, this view rejects every kind of human mirroring in the therianthropic being.

[4]For "biophilia" see glossary.

[5]The father of socio-biology Edward Wilson speaks of the aesthesic charm of the non-human animal, which he generally defines "biophilia". According to the author, such feeling can adjust to the phylogenetic expectations of human beings. We are talking about an a priori aesthetics, which might seem to be a risky hypothesis: it amounts to affirming the existence of an innate perceptive Gestalt as well as a taste orientation preceding any experience of the subject. However, if we consider aesthetics to be one of the many forms of selection, it might seen as part of the context of key-signals. Therefore for Wilson, just like a key-signal, the animal form can please human senses arousing stupor and admiration. See E. Wilson, *Biophilia*, Harvard University Press; New Ed edition, 1990.

[6]For "synmortphy" see glossary.

Indeed, the human being mirrors and recognizes itself in the non-human animal, even if the animal mirror theriomorphizes it and breaks its narcissistic repetitiveness: it gives back a modified image, one that goes beyond my own shape. Animal epiphany is precisely this non-narcissistic mirroring, which, one the one hand, allows for the birth of new existential dimensions and, on the other, gives way to a process of doubling that allows the human being to see itself. In order to do this—as Plessner rightly has it—we have to go beyond ourselves. However, to recognize ourselves in the heterospecific being, we have to recognize our common animal being, and this can be done on many levels: functional congruence, taxonomic resemblance, immersion into aesthesis, vulnerability. Diversity is like fog: it is thick at a distance, but dissipates when you get close.

There are also points where this fog does not cover the landscape, but rather gives shape to it: it helps me better understand myself. This is what happens when we encounter a mammal, with its epimeletic endowments coming from our shared parental archetype. A mammal playing represents a need of experiential apprenticeship to define a canon of belonging that is not phylogenetically granted[7]; its individual flexibility is given by a cortex able to accumulate experiential data. Its strong social investment makes it very close to us, because it engages in organized social interaction even on a very vast territory.

I recognize myself in the youthful shape of a cub because we belong to the same evolutionary model: with its rounded front, flat snout and big eyes,[8] it spontaneously elicits my adoptive disposition. In that feeling I find my authentic and profound qualities: I understand that love has nothing to do with reasoning, it is not algid nor pure, rather it is as opaque as the body, as organic as faeces, as immediate as tears or hunger. I find my humanity in the animal that therefore I am.

I have to get lost in the fog to find myself again. At this point, something unexpected takes place, frightening and astonishing me: the *thaumazein* grounding my research. The animal that therefore I am is the sublime; it is Leopardi's infinity into which I can project myself. Therefore, animal otherness is not a closed door, unless—as Nagel does[9]—we close the door before the encounter even takes place. Heterospecific difference is a train heading into the unknown: it takes me on new roads and calms me down when the thought of the animal that I am becomes too frightening. Animal difference slowly loses its unfamiliarity and becomes a space of

[7]G. Bateson, *Steps to an Ecology of Mind*, Jason Aronson Inc., Northvale, New Jersey, London, 1987. The English anthropologist affirms that play is a feature of all mammals, representing a relational experience predisposing a meta-communication where body movements allow to convey precise information. This is a very significant process as it enables the agents involved to exchange knowledge.

[8]Konrad Lorenz speaks of *Kindchenschema* (Baby Schema) to designate the series of neotenic morphological characters of mammals, such as big eyes, round face, pronounced forehead, chubby cheeks, able to arouse interspecific epimeletic instincts. See K. Lorenz, *The Foundations of Ethology*, Springer 1981.

[9]T. Nagel, *What is it Like to be a Bat?*

possibility, a new home. The animal other loses its exclusively phenomenal nature and becomes an epiphany: the animal that therefore I could be.

In a bird's flight I can find not only the phenomenon's beautiful and sublime nature, but also an achievable existential dimension, so that I find myself flying with a bird's wings. The bird's dance enters my body, possessing it, and the animal that therefore *I am not* moves my body in ways I did not know were possible. To watch a bird fly does not only mean to admire its movements or to learn how to fly; it is first of all inspiration: to learn that flying is possible. The moment I let myself be possessed by the animal that therefore I am not, difference is no longer distance.

I have chosen to follow these three paths of closeness to underline how the humanistic reading of the animal as foreign—(1) foreign because I cannot find any communality to it, (2) foreign as alien, incomprehensible and unreachable because hidden into a bubble to which I have no access, (3) foreign inside of me, because I ideologically project myself into another dimension that is overtly distant from animal-being—inevitably leads to formulate the ultimate expression of speciesism,[10] which identifies the animal with a *res extensa.*

Descartes therefore is not an accident along the humanist path, rather: he is its logical consequence. All philosophers who, while remaining within humanism, aphoristically express their closeness to non-human animals, do not realize the inconsistency that undermines their albeit praiseworthy propositions.[11] First Descartes and then Heidegger, more than any other author, have coherently expressed what is implicit in humanism, namely, that the human being is not "an animal endowed with reason or language", but more simply it is not an animal.

3.1 In Search of the Animal

Proclaiming our Promethean ancestry and stigmatising Epimetheanism as the dimension of need, we inevitably ground the magnificent and progressive fate of the human race in the dimensional disjunction that ultimately divides the human being from the non-human animal. We fall back into the fog with the feeling of not having a definite profile. By refusing heterospecificity, we turn animality into our horrific neighbour. It becomes a vestige of which the human being has to cleanse itself, and animal self-recognition becomes a synonym for ancestrality: closeness to animals takes on a significance of contamination and obscenity, losing any shamanic-progressive connotation and turning into regression.

[10]For "speciesism" see glossary.

[11]Leonardo Caffo defines these forms of humanist speciesism "crypto-speciesisms". L. Caffo, *Il maiale non fa la rivoluzione. Manifesto per un antispecismo debole*, Sonda, Casale Monferrato, 2013. The productivity of these forms within the twentieth century discussion of animality is undoubted. However, we cannot call animalism a real criticism to speciesism, as it rests on speciesisms (indeed, cryptospeciesisms) hidden by the liberating intent but still invalidating in terms of marginalization and deprivation of meaning.

Humanism implies that we need to establish a gap between us and animality, declaring our non-belonging to the animal condition. However, if, with intellectual honesty, we scrutinize the ontic disjunctions that we try to hypostatize, their tautological nature will become manifest. If for a brief moment we move away from the humanistic boundaries that make us consider ourselves as non-declined entities —i.e. not conformed to adaptive functions—dismantling our claim of metafunctional neutrality, we'll find the animal other even in deep ethographic difference and remote phylogenetic connection. Animal otherness talks to us and makes itself understood, at least on a basic level that we share with it at the very heart of our humanity. When approaching a hornet's nest I can recognize its menacing movements, just as I can understand when a cat is frightened, or the courtship dances of a bird of paradise, the movements of a mosquito to avoid my hand, the worries of a female for her puppies, the whimpers of an injured dog. The unknowable animal is a Humanistic invention: *Umwelten* are not separate dimensions, but overlapping contexts that have many meeting points.[12]

Chimpanzees and Gorillas that have learnt a new model of communication in order to dialogue with us—which, by the way, is what dogs usually do—have gone much further than Nagel. Of course, we cannot take everything for granted, since translations are often full of mistakes due to the extremely common "false friends" we encounter in trans-specific dialogues. However, someone may object that Nagel may have wanted to say something different, more profound, when speaking of effect. Hence the question: what does "effect" (i.e. being a bat or feeling like a bat) mean? These words cannot be trivialized, as many philosophers tend to do, by referring to them as perception or even sensory immersion. Being/feeling like a bat involves positionality, which itself involves perceptive immersion and is not subsumed by it.

Animal being means being "referred-to", i.e. having an intentionality that connects you to the world in a particular way. Even before the senses, emotions and motivations define the subject's connection. If we place a cat and a small kid on the grass, the former will pick flowers while the latter will probably chase a butterfly. The grass is more or less the same for both, from the point of view of perceptive immersion (between humans and cats there is not the same sensory difference that you find, for instance, between humans and bees), but intentionality, namely the connection to the world, is of a different kind. While the human is attracted by what

[12]As I have underlined in *Il concetto di soglia*, Theoria, Roma, 1996, the human/non-human threshold is for humans an issue of hospitality. In that sense, Hestia is the goddess who hosts and asks to be hosted. If we analyse the Latin etymology of the concept of hospitality we'll see that it comes from hospitium: such term designated an institutionalized pact between the leader of a community and the hospes (the host). It is in this light that we should look at non-human otherness: nor as mere tools at our disposal but as autonomous and subjective entities full of performative power. Such acknowledgmenet of the citizenship of non-human otherness is the first step to take towards a double decentrative hospitality: a form of welcoming while being welcomed by the other-with-ourselves.

emerges from the environment, the cat is attracted by what moves in the environment: the perceptive immersion is similar, intentionality changes.

The cat's effect, just as human effect, means being-feeling connected in a specific way, irrespective of the context. On a beach, a kid would probably collect shells while the cat would chase bits of sand moved by the wind. Intentionality does not depend on the context, as it only defines the targets (flowers, shells, butterflies, sand) available to a particular kind of intentionality (picking or chasing), where their being targets is not an absolute characteristic, but it depends on intentionality itself. The intentional character belongs to the individual, namely, it is an inherent character defining the subject's positionality in its interaction to the world. But, then, can the effect be defined as unknowable?

If we humanistically assume that thoughts are the only certain thing—which means that you can doubt the existence of anything apart from thought—then all our speeches can be charged with a naturalistic fallacy. The animal is here, in us and in front of us, and it's a being-for-life: a life that includes death, which is not the simple ceasing of bodily functions, but a real expression of life. The animal that I am and am not is more real than thoughts, it runs through the paths of life and can find happiness; it is a body expressing itself in a Dionysian way, defying the laws of thermodynamics and transforming resources into endowments, laws into spaces of freedom, and time into an internal universe that can be filled with infinity. The non-human animal allows for metamorphic mirroring and doubling: hence it lays at the basis of self-reflective thought.

Who are, then, animals? In general, the word "animal" refers to all that moves autonomously but is not human. They are "around us" as spaces for discussion facilitating signification processes and identity construction.

The image of the animal, in this sense, can be:

(a) polarised, that is, it can be transformed through contrast and antinomy to the human features we want to enhance;
(b) homologated, namely, it undergo a reduction of its multiform and plural signifiers in order to give space to a single category;
(c) stigmatised, to underline danger, deviation, contagion, regression or, in general, to indicate any negative predicate we can find in humans;
(d) assimilated—through predicates that work like metaphors, metonymies, or symbols—to indicate qualities desired by the human being, like loyalty, strength, courage;
(e) disguised, acquiring fantastic meanings and being transformed into a space for wonder, dream, nostalgia or imagination;
(f) reduced to something familiar to humans, for example in antropomorphization or reification.[13]

[13]For "reification" see glossary.

Therefore, what emerges here is never the heterospecific animal, belonging to a specific phylogenetic dimension from which the human being is detached; rather, it is the kind of functional connection that man feels towards non-human beings.

We could affirm that in front of the heterospecific animal we immediately become dazed and poor in world. We should probably speak of animality, and understand that every time we enter this topic, we also enter a recursive flow that does not allow us to have a clear image of the animal. Undoubtedly, this is an inclusive argument for a number of reasons:

1. an epistemological problem, namely, we do not know whether to choose an anthropomorphic or mechanicistic interpretation; whether to follow the path of identification and projection or favour distancing and consider animal expression as any other abiotic phenomenon;
2. a self-reflexive difficulty, i.e., humans feel they have a common ground with other species, as they belong to the condition of animality, so that they feels involved in every discourse on animals, both because they feel continuity (being animals themselves), and because they desires discontinuity (to distinguish themselves);
3. our field of investigation is deeply contaminated, since the human itself has gained traits non-human animal through culture, and that is why it finds human traits in heterospecific beings. Since these characteristics belong to the realm of culture, as they are not present in our phylogenetic heritage, it follows that some human traits are closer to heterospecific beings than to the human being.

Therefore, we should not be surprised if we find ourselves in an inclusive situation made of recursive functions: each animal evaluating interface is built on a previous introjective connection of the animal itself. The non-human animal is close, we have the certainty of its existence, but we struggle to focus on it, because our lenses—human predicates—deform its image, as they are themselves the product of an inclusive relationship to non-human animals. If we accept the principle of animal epiphany we will inevitably fall into recursivity, as the human interface, which relates to non-human animals, is already the bearer of non-human animal content. To escape this entanglement is not easy, but it is necessary to find grounds of reflection, while being aware that every point of view is always partial. It is like walking in the mist, seeing the scenery at intervals, and putting the pieces together so that we finally have a partial, contaminated picture that works as a starting point. We must be aware of the limitations and interpretive biases we may encounter, namely of the Gadamerian hermeneutical circle that complicates the extraction of heterospecificity as such. We must resist disruptive and deconstructive temptations, like Nagel's unconditional surrender or the arbitrary attribution originating in Heidegger's mirror.

In the humanistic framework, the heterospecific being is stripped of its heteromorphism, i.e. of its species-specific dimension, by means of two apparently opposite operations:

1. categorization as the human's opposite realm, that is, general and undifferentiated "animality", where non-human otherness does not raise any cognitive interest since it can be obtained by a simple reversal of the concept of humanity;
2. deconstruction of the idea of species (another arbitrary categorization), i.e. its reduction to a generalized "individual". Here the ontogenetic dimension is subsumed under the identity of a particular individual.

If sealing species boundaries transforms the heterospecific being into a mere phenomenon[14] that can be observed but not encountered, it is also true that its categorization as a general individual does not allow for animal epiphany, as it deprives heteromorphism of its significance. However, this univocal interpretation of other animals is easily avoidable if we think of non-human individuals as:

(a) animal, or trans-specifically linked to the human being, since we share some predicates that can be attributed to the condition of animal-being;
(b) heterospecific, or belonging to a community of individuals that share a specific phylogenetic journey and thus display some common features;
(c) individual, endowed with their own experiential, ontogenetic, biographical dimension—i.e. an identity.

The non-human animal is an individual—that specific dog, cat, or gorilla—while the category-words "cat", "dog" and "gorilla" are not arbitrary inventions, since the "individual being" is a consequence of its "being species-specific". The individuality of that specific cat is a function of the general "cat-being", which is different, for instance, from "dog- being". In other words, the presumption of individual horizontality—the idea that species are a human invention—is a huge mistake, since an individual can be "that cat" only if, first, it is "cat". Therefore, cat-being exists even before the individualization of *that* cat and, indeed, we can state that the individual is that-cat exactly because it performs cat-being in its own specific way.

Species-specific identity is the field of interpretation allowing for the emergence of specific individual identity. If we cancel the former, the latter will inevitably disappear too. Then, what does cat-being mean? Trying to avoid an eidetic reading, as well as the nominalistic idea that considers cat-being as the sum of "individual cats", we can state that species-specific identity is the ontogenetic information that has been diachronically gathered about the taxon.

The individual cat consists, predicate-wise, in a recipe whose information is distributed across various sites, with which its ontogenetic process inevitably communicates:

(a) the genome defines a series of constitutive data in order to realize the basic constitutive material, and it does so in a species-specific way, so that its modifications usually determine a transformation of structural characteristics;
(b) the epigenetic complex, from cytoplasm to uterus, produces more morphogenetic information, not only in the specific genome translation, but especially

[14]As suggested by von Uexküll's concept of *Umwelt* or Niko Tinbergen's concept of ethogram.

in dictating the timing and stages of development that give life to specific forms-functions;
(c) parental care produces further information outside the gene pool, passed down from one generation to the other. This information determines behavioural aspects that otherwise would not develop in the individual and that represent an imprinting that moulds the individual according to the canon of species;
(d) the overall constitution of the organisms—from their senses to their biomechanical, postural and kinaesthetic assets—determines a precise range of possible experiences that are calibrated on each species;
(e) the ecological niche, which is always species-specific, and which, while being subjected to annual fluctuations, presents experiential and stimulus-related regularities favouring the construction of some kind of species canon, produces further homologating evolutionary information.

All this information—sedimented during the course of phylogenesis—defines a species identity with which the individual (that cat) must come to terms when building its own unique identity that represents an interpretation of it. Therefore, abandoning essentialism does not mean giving up species identity, even if we consider the latter as a feeble identity emerging at a precise moment in the history of life on Earth, subjected to constant adjustments according to phylogenetic necessity, and destined to be fragmented into other identities.

To state that "cat" does not exist because what is in front of us is *that* specific cat is a mistake that leads us to lose sight of *that* cat too. Cat-being does not linger outside the Platonic realm of other identities. Cat-being is a specific way of feline-being, which is a specific way of carnivore-being, and so on along the taxonomic chain. Hence, in cat-being we can find:

(a) specific features that differentiate cat-being from feline-being and so on;
(b) common features for which cat-being is not different from lion-being, or, along the chain, gazelle-being.

In this sense, while Platonic fixism gives us species identities consisting of non-overlapping monads, in the Darwinian evolutionary idea the different species identities present overlapping areas as well as areas of specificity.

It is quite clear that if we believe that heteromorphism does not interfere with human identity, we will never understand the concept of animal epiphany. With respect to this, we are faced with three different interpretations of the heterospecific individual, none of which is free from error:

(a) projective anthropomorphism, where the human being only takes the principle of sharing into consideration, transforming human identity predicates into universal features and projecting them onto the heterospecific being, reading the areas of non-overlapping as shortcomings in order to affirm quantitative continuity or discontinuity;
(b) functional essentialism, where the human being only takes the principle of specificity into account, understood as the species' essence, reaching two

opposite interpretive conclusions: the monadologic idea of species, implying animal unknowability, and the application of a mechanist model to interpret adaptive features;

(c) inclusive categorization, where the human being neglects the emergence of identity along the whole taxonomic chain, reaching two contradicting conclusions: the individual's desubjectification, where the individual becomes a mere specimen of its species without any interpretive possibility here-and-now, and the transformation of all heterospecific beings into a single animal category opposed to the human.

These are three erroneous interpretations of non-human individuality, all of which find a synthesis in von Uexküll's concept of *Umwelt*. Therefore, we need to question this paradigm, which has had a great influence on philosophical thought also thanks to its clear reference to Plato.

3.2 The *Umwelt* as the Animal Prison

Following von Uexküll's monadological idea, two things can be stated:

1. the animal, closed in a predetermined bubble, where perception and action are the same and every action has been established, cannot interact with the object and, consequently, cannot emerge as a subject;
2. the various *Umwelten* are separated realities, so we cannot tell how being a tick, a bee, a bat or a lion feels—using the same animal examples as those by von Uexküll, Heidegger, Nagel and Wittgenstein. Hence the paradox in von Uexküll's essay: as nothing can be said, nothing must be said (since if you cannot say anything, you must say nothing).[15]

In my opinion, these statements are both erroneous, since:

1. animals are the protagonists of their own lives—which means they are subjects —despite having a specific perspective on the world, which however does not mean they are trapped in the cage of their own vital needs;
2. it is not true that human beings cannot know what a heterospecific being feels, since, although we cannot reach their most intimate and detailed feelings (which is also true for other humans and even for ourselves), we can reach a good level of approximation using the Darwinian paradigm.

I'll start from the last point. If we accept the basic idea of Darwinism, that is, that living things have developed through millions of years starting from common

[15]Following Wittgenstein's statement that: "What can be said at all can be said clearly; and whereof one cannot speak thereof one must be silent". L. Wittgenstein, *Tractatus logico-philosophicus*, project Gutenberg 2010, p. 23.

progenitors, it follows that there are some common traits among different species, based on three coordinates:

(a) universals, or common features among species that respond to non-specific or general needs like the ability to feel pain and get away from its source, sensibilization, the tendency to repeat actions that cause pleasure, and perceiving the world according to patterns (Gestalt) that can bring out figures.
(b) homologies, or common features between two species descending from the same ancestor; in this case the species will have the more common features the closer they are from a phylogenetic point of view;
(c) analogies, or specific common features originating from the species being forged by the same selective pressures, especially concerning the environment, lifestyle, reproduction, social structure, and food supply.

These three coordinates suggest that the *Umwelten*, if we can still call them so, are not isolated monads, but they rather present overlapping areas whose dimension increases:

1. the more we consider their basic fitness features, as their adherence to universal aspects will grow;
2. the closer they are from a taxonomic point of view, as they will show more features inherited from a common ancestor;
3. the more analogies they have regarding the species' lifestyle, habitat, reproduction, sociality, and diet.

In this respect, we can represent the spheres as partially overlapping entities whose shared space indicates an area of living commonality. This space, which I call "ontic sympathy", will be the more extended the more responsive the comparative event is to the three similarity parameters (according to species or character).

When a fish fidgets because it has been hit by a harpoon I can state that it is suffering without being accused of projection, since this is a universal character (1); when a chimpanzee is with other dominating individuals and pretends not to know where he has hidden a banana so that it can eat it alone, I can understand what he is feeling because our taxonomic proximity makes me think we might have inherited a character from the same ancestor (2); when a dog starts a coordinated operative strategy I can understand his reasoning, as the collaborative activities of man and wolf have many analogical structures in common, since, when playing together in a team, some predicates are forged by the same selective pressure (3).

It is clear that if the human being wants to explain the behaviour of heterospecific beings, he has to do one of the following:

1. state his "interpretive impossibility" and apply a complete suspension of any form of evaluation—this is the choice of those who see *Umwelten* as separate monads;
2. treat the animal like any other abiotic phenomenon, as if it were an avalanche, a river or a hurricane, that is, approach animal behaviour through a mechanistic explanatory model;

3. consider the heterospecific being to be identical to ourselves, in a "trivial or projective anthropomorphism" that is applied with no species or predicate distinction;
4. consider the heterospecific being to be a mix of similarities and differences, i.e., apply a "critical anthropomorphism" according to the universal-homological-analogical specificity of that particular situation, in order to face difference according to its specific functional-adaptive meaning, placing fitness at the core of this explanation.

My point of view corresponds to the fourth interpretation. Indeed, Darwinism shows us that the great catalogue of animal species has been realised through "continuism", or the continuity between living beings—i.e. the big areas of animal similarity. However, at the same time, every species has interpreted the similarity principles and fitness postulates in its own specific way. Therefore, anthropomorphising certain species behaviours does not represent a contradiction of Darwin's principle; also, every form of anthropomorphism has to be evaluated on the basis of solid scientific knowledge of the species on a taxonomic, ecological and ethographic level.

Critical anthropomorphism is based on two considerations:

1. in order to evaluate animal predicates we need a comparative element, which is to be found in humans rather than in machines;
2. when speaking of comparison, we do not refer to a projection or translation of predicates from man to animal, rather, we evaluate the spaces of similarity in order to better understand the differences. The space of sympathy—which is not interpreted as "feeling affection for", but as "feeling harmony with"—has to be considered according to the following points:
   (a) character;
   (b) species.

The space of sympathy, where anthropomorphism is maintainable, becomes an epistemological pull to postulate and test explanations. Critical anthropomorphism, therefore, does not consider the anthropomorphic explanation as final, but as a starting point that can be falsified, and one that aims at showing difference, as is customary in comparative anatomy. Indeed, critical anthropomorphism, or the space of sympathy, highlights the areas of difference or specificity. This is because every predicate responds to:

1. the functional principle, i.e. those shared needs (feel pleasure, reproduce, avoid pain, express dispositions) characterising our belonging to the animal realm;
2. adaptive specificities, or species declinations, which can be deduced through the analysis of the species' anatomical, physiological, behavioural, ecological structure.

The analysis of difference is achieved starting from the common ground of our shared animal-being, through a method that places biological needs at the core, with functional specialization as their declinations: we can call this approach "functional biocentrism". If, for instance, I know that eating is a necessity of living creatures and that every necessity is physiologically marked with pleasure, I can imagine how a bee feels when eating nectar or a tick when sucking blood: I know it feels pleasure. Similarly, if I am aware that every animal feels gratification and fulfilment in expressing its motivational dispositions; I can imagine why a dog experiences ecstasy when sniffing around, although I do not relate to that as a human being. In other words, through functional biocentrism I can slowly move from the space of sympathy towards those areas where empathy is needed because there is difference.

Functional biocentrism is the method normally used by ethological research. For example, we can suppose that social animals like wolves or dogs use some communication signals to regulate interactive behaviour—such as warning, engagement, hostility, assertiveness, pacification, submission. From this space of sympathy, where animal living is central (if you are a social animal, you need to know how to relate to the other group members, and communication is much more efficient than interaction), I can start my research on a species' zoosemiotic specificity.

Biocentrism is used not only by ethologists, but also by the very dog owners who have seen their dog being inebriated by the smell of their shoes. Biocentrism consists in recognizing that the heterospecific being has similar reaction to us, even though they are expressed through other variations. Basically, this means that I recognize a satisfied orientation in the dog's perceptive action—as demonstrated by the fact that its olfactory expression clearly shows the signs of pleasure and gratification, rather than those of research. Indeed, the dog:

(a) is not looking for anything on our feet;
(b) while smelling/sniffing, it starts being more relaxed and excited; in other words, it gives rise to dispositional involvements—which however respond to a different aesthetics in terms of both sensory priorities and sensory taste.

To put it differently, life has its own coordinates of pleasure, pain, gratification, frustration and so on, which always have a central role. We never stop being bodies, because our mental life is always the outcome of a combination of physiological factors: we reason with our glands, with the cytokine flows of our immune status, with metabolic differences and the overall biorhythms of our whole organism. We are bodies when we believe that we possess a body and even when we completely forget about it. Duality is only apparent: it results from an ideology of separation and distancing, hierarchy and exploitation. Life is always central, even when we think we are-towards-death, since death is not a cessation nor a denial of life, but an act of life. Being a body makes all animals incredibly alike, as in all of them life is central, with its vulnerability, its non-equilibrium, its dependence. Being a body means that projections beyond the body—in Plessner's ex-centricity—are nothing more than expressions of the body.

Critical anthropomorphism is only a step towards a more mature interpretive ability, comparable to a child who recognizes the difference between his mother and himself. The biocentric[16] interpretation allows for the emergence of difference, which, in this perspective, is no longer cause for incommunicability, since biocentric similarity represents the Rosetta Stone that makes translation possible. Our shared being-a-body becomes our common watermark, which does not nullify differences, but clarifies them by viewing them from an interpretable perspective. Difference becomes the way through which each species reaches biocentric coordinates by different paths. Biocentrism means recognizing each other through our difference, since we do not need to be identical in order to "feel together", that is, to understand what the other is feeling. Not even identity is needed to feel the other's coordinates in our own body, acting on our chords. The monadological view, which leads to a sceptical attitude of superiority, hides the need to deny nonhuman animals any authority. There will always be someone stating that "after all there is no scientific proof that animals feel pain." And indeed, if Nagel is right, we will never know, since investigative denial is already present in the rhetorical content of the question.

Let us consider our first point: the idea that the *Umwelt* is a predetermination of life and a negation of subjectivity. Indeed, if the animal only consisted in its automatic reactions to a predetermined set of stimuli, its space of subjectivity would not exist. Reaction or feedback is not properly an action; it is not even an interaction, let alone a relationship. Between a tick coming off a branch if stimulated by butyric acid and an engine starting when I turn the key there would not be any difference. In this reading, von Uexküll limits himself to reworking the Cartesian idea, enriching it with perceptive and operational detail, but still dealing with automatism, since:

(a) the animal cannot say "no" to its marking;
(b) the marking attaches to it a function to perform, which means that we are not speaking of perception, as the animal does not perceive anything but only reacts to stimuli;
(c) reaction is no answer, as the animal cannot help following prefixed mechanisms;
(d) we cannot talk of a real individual identity, as the animal lives in an eternal present, completely heedless in its here-and-now with no memory or future projectuality;
(e) the continuing flow of environmental marking and regimented impulses does not allow it to think, wait, get bored, or (especially) to be self-aware.

This is an erroneous reading of phylogenetic endowments, which are not considered as predicates at the disposal of the subject and at the service of its individuality; rather, they are taken to be strings that, in an imperative and binding way, impose certain behaviours upon it. "That cat" has not decided to be a cat, therefore it has not chosen the terminals of ontogenetic information it has received during its development—from conception, embryogenesis and parental care to its specific

[16]For "biocentrism" see the glossary.

openness to reality influencing its experience in a specific environment. However, once that cat has absorbed this information, it creates its individual heritage, just like direct experience. In other words, the animal becomes the owner of its endowments, which play the role of instruments or expertise it can freely and creatively use. If it were otherwise, the individual would stop at the first inconsistency in its here-and-now—and inconsistency is no exception, but ordinary routine, since the animal lives in a complex reality where there are so many intervening variables that there can never be two equal states of reality. Therefore, the animal must be highly flexible, which means it has to act and not just react to the world, as this would determine its failure. The world is, in fact, in a permanent state of chaos. The animal has to seize information—be it phylogenetic heritage or individually obtained information—and treat it not as an algorithm, rather as "building material", namely, solutions (heuristics).

The animal's ownership of its endowments allows for the emergence of a necessary predicate, i.e. subjectivity, which should not be interpreted as *regressio* ad infinitum, or as a hidden entity inside the machine, but rather as something originating in the structure of endowments, in their internal plasticity (co-optability to more situations and functions) and in their multiform compositionality. If every species lives in a context-world and presents specific operative predicates, it is also true that these cannot have a rigid functional structure, because they would not serve the adaptive purpose. Hence, the subject does not only follow environment marking or drives: it reflects its here-and-now following its own coordinates, which are unpredictable as they always emerge in a present singularity.

That is why we can state that an animal is *endowed with* a here-and-now and is never *included in* a here-and-now, as its memories and projects allow it to overcome the present and integrate it into a diachronic flow. We cannot break the diachronic flow and interpret the subject's here-and-now: the past gives shape to the future and the projection into the future gives a specific and functional meaning to the past. The individual must always maintain ownership over its endowments and remain in the flow—avoiding being comprised in the here-and-now—if it wants to engage variability.

Creativity, even though expressed in different ways by various species, is the real predicate of animality, understood as "subjective presence in the world", or what allows individual self-realization to take place starting from sedimented endowments. However, we should not consider subjectivity as individual autonomy or the individual's elevation above the world. Creativity is not an expression of detachment but of love and wonder, rapture, engagement with the world. Individual creativity is driven by languor and tension, openness to the world; it is never fixity in the here-and-now, but rather excitement, and the incapability of being entirely positioned in the here-and-now.

That is why the idea of a perceptive-operative delimitation of animal behaviour is misleading. When we consider the lives of animals with which we do not have much in common, and thus are very distant from us—as is the case with von Uexküll's tick or Heidegger's bee—our ability to observe individual detail is reduced. In fact, as close as we can get spatially, between a human and a tick or a bee there will always be a deep philological distance. There is no way for us to get closer: we lose sight of detail, and we end up homologating its behaviour. A dog is closer to us and a chimpanzee even closer—a few million years is a short distance —and this allows us to see the detail of their subjectivity.

However, be they distant or close, animals respond to the principle of creativity, so that in their case we can apply Heidegger's concept of *Dasein*, since animal-being means being self-aware, world-creators, relying on one's endowments to let one's own individuality come to the surface. The endowments are the colours and the individual subject is the painting: you cannot paint without colours as you cannot identify a painting with its colours. Subjectivity is the artist who can create an individual profile, a coherent painting, from different colours. When speaking of a subjectivity with endowments, we are not referring to nature's mannerism; rather, to a necessity: in fact, preformed paintings or a colourless artist would be disadaptive.

3.3 Animal Appeal

It is quite clear that animal otherness represents a gravitational force for the human being, but this simple truth is not always openly stated, perhaps due to a sort of anthropocentric modesty that requires us not to be exposed to non-human fascination. The other species' charismatic presence seems to be an escape for human imagination, which can thus discover unexplored territories. Animal fascination reminds of rapture and bliss: the annunciation of a possible space of growth. And if our discretion, when not direct denial, in dealing with the debts we have towards other species reveals an autopoietic unrealistic ambition, proofs of hybridizations abound.

Children are immediately enchanted by the animal world, almost testifying to the innate biophilia expressed by Edward Wlison.[17] Adults must follow this electivity if they want to capture children's attention: thus fairy tales, toys, and cartoons become animals. The bond with animals lies deep in the human being and the child's self is able to bring it out. Anthropologic research is so rich of "zoomorphic" references (visibly linked to a particular species) or therianthropic references (linked to a nonhuman but non-specific condition), that Claude Lévi-Strauss had to admit that there must be a very strong connection between animals and the categories of the human mind.[18] The animal "good to think"[19] challenges the purely reifying vision

[17]E. Wilson, *Biophilia.*

[18]C. Lévi-Strauss, *Structural Anthropology*, Basic Books, 1963.

[19]Such expression, proposed by Lévi-Strauss, can be ascribed to his studies on totems, where the animal catalogue has always been a preferential element (a sort of compass) to orientate through nature. Animals, in this reading, are very useful tools for the processes of human symbolization (to all effects, then, elements that are "good to think") since with the plurality of their forms they offer

that has become a leitmotiv of the anthropological tradition, breaching the isolationist idea of humanistic anthropocentrism.

Undoubtedly, in Levi Strauss' proposal there is also an instrumental resonance, be it only for the preposition used ("to") that desubjectifies the heterospecific being turning it into material with which to "dress up" thoughts that were already present in the human mind. However, there is also evidence of an intersubjective relationship—encounter, dialogue, confrontation or transaction—based on content flowing from non-human to human beings and vice versa. Its manifestation as an aesthetic incipit, with zoomorphism being the first artistic canon, clearly shows that inspiration starts from the animal, which is more that mere disposable material: it is the starting point of a revelation. Art itself is man becoming animal: trans-specific metamorphosis. The heterospecific being is not a figurative-narrative theme or *topos*; it is not content but container, turning human into a transhuman being, creating a synthesis of the beautiful and the sublime in therianthropy.

The first Palaeolithic drawings—or animalist art—represented heterospecific beings. We can find the obsessive presence of animal forms in all Neolithic objects, where the heterospecific being is a living presence, be it iconic or propitious. But not just that: art is first of all hybridization: morphopoiesis of werewolves, harpies and sirens, one that can start new ontological biomes, where the achievement of human condition is always a Mephistophelean pact with the non-human animal—an overcoming of our phylogenetic condition. Together with sexual organs and human forms, animals stand out for their continuous presence in ornaments and votive statues. Frequently, each animal corresponds to a virtue, so that the object not only has an ergonomic-functional meaning, but it is also charged with the content extracted from the animal it represents. Zooplastic speaks of an inspiring principle of heterospecific representatives, which allows the human being to build a series of objects that reproduce biodiversity in a fractalic way. The human being conceives of cosmopolis as a mirroring of the surrounding nature. Human culture, therefore, before being Promethean, finds its own architecture among heterospecific predicates: in other words, it is Epimethean. This theriomorphization of the anthroposphere[20] is not limited to objects but involves the project of the body. Animal mirroring becomes a principle of identity, ethnicity and personality. Anthropologic research shows that some populations initiate young men into adulthood through zoomimetic actions—tattoos and scarification, neck rings, hats and ornaments, body painting and cosmetics. Animal possession becomes patent in dances representing courting rites as well as in zoomimicry, which is the first form of narration, broadening the human phonetic spectrum, in rituals that evoke other species' postures, gestures, and rhythms, or in music that reproduces nonhuman harmonies, like the very first models of flutes and percussions.

(Footnote 19 continued)

a rich (potentially infinite) system of differentiation, that can activate a fruitful process of classification of reality and society.

[20]For "anthroposphere" see the glossary.

We are dealing with a real "animal appeal"[21] where the other summons and decentralizes the human being, taking it into hybrid existential territories like the one of possession, where the human body moves along non-human expressive coordinates. The animal is, therefore, an epiphany, as it overcomes the mere appearance of its presence as phenomenon and brings the announcement of a different existential dimension. This means that heterospecific fascination is not an end in itself, but it rather determines a decentralising gravitation and the revelation of other existential possibilities. For us, watching a bird fly does not mean we understand flying techniques, but it suggests the possibility of flight. In other words, it opens a dimension onto which man can project himself first through imagination, and then by planning a way to do it himself. Without this inspiration—that is, without "seeing himself in the flight of a bird"—no mastery would be possible. We should consider this decentralising presence that, similarly to the tinfoil origami in *Blade Runner*, has an erosive effect on our castle of anthropopoietic certainties, which has always been man's solipsistic emanation. Heterospecific epiphany shows us that we cannot understand man if we enucleate him from its relationship with nonhuman animals. This epiphany marks the beginning of a new philosophy that replaces the Cartesian *cogito ergo sum* with *dialogo ergo sum*, acknowledging the existence of a dialogic situation with nonhuman beings and its contribution in the realization of the human being. In this sense, we can speak of a posthuman approach, as what is being rejected is the humanistic idea that founded the human being *iuxta propria principia*.

It is fundamental that we understand whether the non-human animal somehow appeals to man and, if so, how. Of course, we cannot identify a single gravitational-orientative source or look for an explanation that involves the *imprimatur* of completeness. The animal appeal originates from a substantial number of motivations affecting different aspects of human orientation, so that establishing the prevalence or priority of one or more motivations over the others would be difficult. Moreover, cultural hybridization with other species, or the animal transformation operating in dance, cosmetics, music, and so on, leads to human dependence on the external referent and, thus, to an orientation towards heterospecificity. Such an autocatalytic effect—or the positive effect created by hybridizing properties[22] according to which for every animal-becoming event there

[21]R. Marchesini, K. Andersen, *Animal Appeal. Uno studio sul teriomorfismo*, Hybris, Bologna, 2003.

[22]Elsewhere I have defined this process as "zoothrophy" in the sense of a typical tendency of our species to "turn to other animals" and recognize them as social counterparts, as well as choose them in anthropopoietic processes. This word indicates a clear distancing from other non- or proto-zooanthropological explicative structures, i.e. not founded on a dialogic-referential role of the heterospecific being. The theory of zootrophy operates an important epistemological shift as it considers the heterospecific being interesting not as a stimulating element—be it for its phenomenic richness, for its role as an elicited response to many reasons, or for its identifying potentialities as living and recognizable being—but as an interlocutor with referentiality that can bring about new contents through dialogic action. R. Marchesini, S. Tonutti, *Manuale di zooantropologia*, Meltemi, Roma, 2007.

is a human dimension that is more prone to therianthropic fascination and more incline to hybridization—is ascribable to a phenomenon of self-perception or self-efficacy. For instance, when the human being introjects the animal through dance, dancing becomes a human dimension and the human appeals to non-human kinaesthesia by searching, in a deeper and more exemplary way, for its own condition, now decentred from the phylogenetic legacy.

Therefore, to look for the *leitmotif* of animal appeal only among phylogenetic characteristics would be misleading, since what we would find in terms of heteronomic openness is never a simple ethographic structure, but rather an anthropopoietic construction site. However, a brief description of this phenomenon and an explanation of zootrophic motivations might be useful here, as they allow us to keep track of the most common dimensions in terms of orientation, relation, and hybridization with animal otherness.

Let's start with some purely descriptive considerations. Psychological research shows, through various preferential acquisition tests, that there is a perceptual election of zoo/therianthropic targets. During the test "find the hidden object", a research carried out by Paul Shepard[23] and myself, we have observed that:

(a) within a defined period of time, the animal target is found at about 90% whereas the target attributable to common objects is found at about half;
(b) within a non-definite period of time, the animal target is detected in a much smaller amount of time, around 40% less than other targets.

More tachistoscope tests show that in a collection of targets and stimuli, people tend to remember the animal target: hence the hypothesis that zoomorphism can be considered a prototypical stimulus, like other semiologies (for instance the colour "red"). Another test concerning the attention span shows that, again, the zoomorpheme[24] positions itself next to stimuli that are highly correlated to attentive orientation, such as something unexpected, new, or sexual characters. Even ethological observation shows that the human being, regardless of the specific culture, tends to give an animal shape to uncertain morphologies. This is how constellations give rise to a zodiac, clouds become sheep, a mountain profile turns into a horse or a dog's face, a combination of intricate and twisted lines reveals an animal figure and so forth. Shall we think, then, of a zoomorphic Gestalt? So it would seem. We tend not only to better capture the animal form than that of other targets, but also to zoomorphise what is uncertain to us.[25]

[23]P. Shepard, *The Others. How Animals made us Human*, Island Press/Shearwater Books, 1996.

[24]For "zoomorpheme" see glossary.

[25]Paul Shepard ascribes this tendency to the hunting nature of man, i.e. to the importance of seeing the animal as a potential prey. Personally, I am convinced of the contrary. Predators are oriented towards a moving target and can better detect the object when it's in kinetic rather than static conditions. Proof shows the opposite: there's a perceptive electivity of static forms immersed into a complex framework, such as for instance the need of a prey.

Our interest in the animal can be explained with reference to two important motivational coordinates of the human being:

1. the sillegic tendency, that is, the propensity of gatherers to collect and catalogue; in other words, based on their diet (which is based on very different targets emerging from the context) they tend to proceed to a distinctive and categorical recognition of the context around them, therefore being very interested in the lay of the multiplicity;
2. the epimeletic tendency, that is, the propensity to recognize young specimen and to be sensitive to et-epimeletic appeals or calls for help and protection. Such disposition has evolved in the human being as young humans need an articulated system of parental attention and are characterised by a long developmental process, so their adults have the necessity to maintain that disposition after the weaning period. When I speak of motivational structures I refer to innate dispositions or phylogenetic heritage manifesting itself as:
   (a) sensitivity towards particular targets, such as multifariousness for sillegy and paedomorphy for epimelesis;
   (b) competence in terms of innate predicates specialized for these activities, both in perception and operation, with the result that many behavioural tendencies are inhibited;
   (c) gratification when performing such actions, namely, the intersection between motivational expression and gratification mechanisms, also in terms of pleasure and positive emotions.

These two motivational structures are at the centre of the fitness dictatum as they are connected to basic needs (survival and reproduction) of mankind. But what do they produce with regards to human openness to the non-human animal?

First of all, I would like to specify that these motivations have an internal explanation, which means they are internally adaptive to species-specific interests and that only indirectly—i.e. as epiphenomenon and collateral effect—they produce openness to heterospecific beings. Under the syllegic[26] profile, it is clear that animals interest humans because they are classifiable—we can even state that the animal universe represents the archetype of every possible collection—and because through heterospecificity we can classify the world in environments, canons, styles, functions, predicates. Under the epimeletic profile, to have a strong motivation means to be very receptive to paedomorphism, to the point that we also take care of other species' babies, with the result that trans-specific adoption, mothering[27] and

[26]For "syllegy" see glossary.

[27]Despite the fact that, historically, there have only been a few studies on this "taboo subject", there are many ethnographic proofs of animal adoption in human communities. These phenomena, together with parental care such as nurturing, breastfeeding, and so on, represent an example of the overcoming of species-specific boundaries isolating humans from heterospecific beings. S. Tonutti, "Il maternage, fra attenzioni parentali e utilizzazione dell'animale", in *La Ricerca Folklorica*, n. 60, Grafo, Brescia, 2009, pp. 30–36.

the inclusion of the heterospecific being into the community are not occasional, but rather—as James Serpell has observed—they are the norm.[28]

If we consider human ontogenetic characteristics, we can understand that heterospecific beings play an important role in the human community. Children develop through interaction with an external referent (usually a parent or another human being) that acts on the child's developmental coordinates in three different ways:

1. acting as a secure base, as observed by Bowlby,[29] namely, favouring certain experiences instead of others on the basis of their safety, thus operating an experiential and evolutive differential upon the child;
2. defining an area of proximal growth—in Vygotsky's words—acting as a model to be reached in terms of target orientation, interpretation of situations, presentation of models, thereby defining a range of identity development in children, i.e. a collection of specific characteristics;
3. giving the child a precise expressive space, in other words, acting as the structure that connects it to the world, favouring certain expressions and inhibiting others, and thus determining developmental paths, since the very fact of expressing a predicate means developing it.

Clearly, the presence of a heterospecific being in a human group necessarily acts in a referential way, since:

1. it is a secure base but it also brings different orientations and behaviours, so it arises new experiential centripetations;
2. it creates a space for heterologous proximal growth as, through social learning, it transfers non-human inclinations, interpretations and operational models onto the child, who reinterprets them in its own phylogenetic dimension;
3. it modifies the child's expressive space, leading it to perform new activities and so changing developmental paths and hybridising the subject's identity.

Last but not least, the human being is fascinated by animal otherness because of the former's marked mimetic tendency. We are used to representing primates as strongly imitative animals, but actually the human being has an ever-stronger mimetic tendency. If a child and a young chimpanzee grow up together, the child will be more likely to learn to act like a chimpanzee rather than the other way round. If we place a human being in front of an anthropomorphic other, again human mimicry is more frequent than its opposite.[30] Children inclination to "play everything" is so strong that Marcel Jousse thinks this is the core of the evolutive process, pointing out that the openness to the world characterising mimetic

[28]J. Serpell, *In the Company of Animals. A Study of Human-Animal Relationships*, Basil Blackwell, Oxford, 1988.

[29]J. Bowlby, *Attachment. Attachment and Loss*, New York: Basic Books, 1969.

[30]E. Visalberghi, *Etologo. Osservare il comportamento degli animali*, Zanichelli, Bologna, 2006.

expression is never solely directed to other human beings.[31] Jousse underlines that when reenacting reality with its body, the human being assimilates it, makes it its own, and transforms it in a "propositional gesture". For Jousse, the unit of measure of language is not the word, but rather the gesture.[32]

Beyond the *querelle* on whether language has its origin in gestures or voice (or, as seems more likely, in a synthesis of both)[33] we should be aware of the particular mimetic aptitude of the human being, which also has a strong competitive element and a certain dose of envy—the "sad passion", in Elena Pulcini's words,[34] that originates comparison. What can be more fitting to mimicry than the non-human predicate? And isn't there envy between the lines of Promethean compensation? Mimesis is a human disposition that cannot be reduced to copying someone else's gesture; rather, it means representing, transforming our body in a theatre that welcomes otherness and interacts with it.

Mimesis transforms biodiversity in an occasion for carnival, a time in which everything is permitted: not only gesture, posture, phonetics or choreograms, but any external element—like feathers or hats—that can make theriomorphic representation more real. Even though all these arguments, if taken separately, would not be enough to justify the astonishing charm that animals have on the human being, however, if taken together they can help us understand not only the various proofs of man's interest in heterospecific beings, but also the natural disposition of children towards them. Animal epiphany is based on this interest, although it cannot be its only condition of existence. In particular, the aforementioned elements are useful if seen in specific mutual relation:

(a) the sillegic motivation together with the mimetic tendency gives rise to an epistemic orientation to the animal;
(b) the sillegic motivation can explain the act of picking up the baby, while the epimeletic motivation explains its adoption and the presence of the heterospecific animal in the human group;
(c) the strong attention to the zoomorpheme could favour warning methods based on animal imitation.

---

[31]M. Jousse, *Anthropologie Du Geste*. Paris: Les Éditions Resma, 1969.

[32]That language has a mimetic nature—despite originating from a vocal interpretation—is clear in Merlin Donald's proposal, however this thesis had already been stated by Charles Darwin in his theory of the three phases of language, according to which language incipit was using human voice to imitate the animal voice. Another author who underlines the mimetic nature of language is Steven Mithen, author of *The Singing Neanderthals: The Origins of Music, Language, Mind and Body*, London: Weidenfeld and Nicolson, 2005. According to Mithen, the first gestural and linguistic signs are to be found in the human tendency to mimic animals.

[33]In fact, if Jousse studies the mimetic tendency of children, Piaget analyses the primary role of echolalia in their communication.

[34]E. Pulcini, *Invidia. La passione triste*, Mulino, Bologna, 2011.

If the human being were not an interested and marvelled observer of animals, we could not speak of epiphany: in order for this kind of transformation to take place, the human being must be possessed by the theriomorpheme. Hence, following Aristotle, we can state that the principle of epiphany is always wonder, and wonder is therianthropic.

Chapter 4
Overcoming the Human

If humanistic thought has placed man as the self-founding centre of humanity, later—starting from Nietzsche's *Übermensch*, who unravels the Dionysian flow and disrupts the crystallization of the form—there has been a slow but inexorable dawn of posthumanist thought that, without falling into easy naturalistic reductionism, promises a new immanent eschatology. The combination with otherness, which goes from Emerson's resonances to Gregory Bateson's connectivity,[1] declines this new attention towards the immanent as a sort of relational ontology that runs through the Nineteenth century, slowly eroding the conceptual constructions of anthropocentric isolationism that had lasted for half a millennium. It is not easy to understand, today, the meaning of an animal that has been impoverished so as to give space to the Vitruvian man. It is difficult to believe in an animal epiphany not deriving from man or God, but rather from a trivialised heterospecific being that has lost its specific predicates and assumed an undifferentiated and distant animal form. This distance is revealing of the way in which man perceives himself and how the non-human being has been paradoxically characterised by opposition to man.

As we have seen, the meta-predicative reading, which should found and mark the dichotomy between man and animal, can have various and often conflicting aspects, which still follow the same tautological structure: man is no longer an animal, as he is human. For instance, disjunctive meta-predicates can:

[1]The research of "the pattern that connects" may represent the highest point in Gregory Bateson's intellectual journey. In *Mind and Nature: A Necessary Unity* (*Advances in Systems Theory, Complexity and the Human Sciences*), Hampton Press 20002, he lays the basis for a real epistemological revolution, in order to find the "pattern that connects" the whole biosphere. For him, the living world is traversed by a series of connections that can link "the crab to the lobster and the orchid to the primrose and all the four of them to me" p. 21. Such relationships are not fixed, internal and constituting the subjects, but they organize the whole biological world through connections fluctuating in time (following a stochastic process).

R. Marchesini, *Over the Human*, Numanities - Arts and Humanities in Progress 4, DOI 10.1007/978-3-319-62581-2_4

1. appeal to an *ab origine* separation, as in the concept of "rankless man", or, on the contrary, be the (*ex-post*) outcome of an ontogenetic process, as in symbolic education and the emergence of the I-that-speaks;
2. refer to a certain quality exclusively possessed by the human being, such as self-consciousness or language, or to man's lack or unsuitability, like his adaptive shortcomings or the low level of specialization of his organs.

It is interesting to understand why man feels this disjunctive need and whether this actually has a *raison d'être*. From my point of view, such a necessity can have many different causes:

(a) a perspective bias, i.e. the one that turns non-Hellenic people into barbarians, increasing the distance between identity and otherness and annihilating the distance among various kinds of otherness;
(b) a categorical need coming from dichotomous-dualistic epistemics, used by man whenever he wants something to stand out and the rest to stay in the background;
(c) an anthropocentric project that has started with Greek philosophy and has been rapidly evolving up until Darwinism;
(d) the ideological justification of the exploitation of non-humans from Palaeolithic extensive butchery to Neolithic socio-economic structures, from the territorial transformation in the Eleventh century to the Seventeenth-century scientific revolution, and from the industrial revolution to biotechnology.

We are not necessarily asked to choose between these two, since we probably find ourselves at the centre of a causal intertwining. On the other hand, I believe we should reflect on how man perceives and represents himself, as I think that this is the key to understand not only our relationship with other species, but also that the human being is not the autarchic outcome of man but rather the result of its relationship with nonhuman otherness. What I would like to investigate here is how even from self-representation we can clearly see that man has not founded humanity in a solipsistic way, but rather through a dialogue with therianthropy. Also for this reason, I believe that the obsessive return to meta-predicative disjunction reveals our relational debts towards nonhuman otherness, and that the polymorphous dimension in which we recognise ourselves originates from this dialogue.

In other words, my point of view is that other animals have prompted in the human the revelation of its humanity, namely, they have opened up existential dimensions that have lead the human to:

1. self-representation, or the ability to reflect on itself and see behind itself, as well as to think of his body as a living being;
2. anthropopoiesis, or the construction of new dimensions of life, imagined and created using the therianthropic structure as an imaginative model of experimentation through zoomorphic actions.

If this is correct, we are faced with a paradox: disjunctive appearance—founded on the human emergence as novelty and ontological uniqueness—is not the outcome of a clarification of anthropological connotations; rather, it is the result of the human conjugation to ontopoietic perspectives inspired by other species.

4.1 Ontopoiesis: Open Identity

Analysing the various expressions of human culture, we cannot help but wonder at the multitude of forms characterising it. Also, it is impossible to catalogue it or trace its typological map, as we come across a continuum of materials and styles where boundaries can only be arbitrary. Such expressive plurality is attributable to particular predicates of the human phylogenetic heritage, such as:

(a) neonatal immaturity, which favours a very flexible ontogenesis that can introject external referents;
(b) our brain's remarkable development especially in its association areas, which transform data into representations;
(c) motivational structures (syllegy, epimelesis, mimesis), which favour the process of conjugation to external referents;
(d) a tendency to externalize functions, namely to use tools—an aspect that was already present in its essence in other anthrophic forms.

Therefore, we cannot give an anthropological explanation without first going through a thorough analysis of these characteristics: just like the other species, man is the result of an evolutionary process that has shaped his anatomical-functional connotations and his expressive inclinations. On the other hand, we should also ask ourselves whether these important characteristics alone (i.e. if they were independent in an evolutionary sense) would explain the cultural multifariousness of man, or whether they would rather need the intervention of external referents to create the human being. To overcome the autopoietic reading we need first of all to discard the idea of man's incompleteness, as we have seen that this lack is easily transformed into an acceleration of distance that lays the basis for ontogenetic autarchy. Overturning Gehlen's hypothesis and with respect to scientific evidence on *Homo sapiens*' heritage, we can presume that the cultural ellipses characterising man's expressive multitude have been created not by a natural deficit, but rather by a certain phylogenetic aptitude.

Such an inclination can be summarized in the three following concepts:

1. a strong propensity to relate to external referents, with the ensuing ability to introject nonhuman beings into the human identity dimension;
2. a strong ontogenetic virtuality, not ascribable to a low supply of predicates, but to a redundancy of evolutionary structures.

In this reading, phylogenetic characteristics are not lacking, but rather substantial, and they cannot be interpreted as deterministic elements of the human being, nor as general virtualities potentially fostering any outcome. Rather, they are factors favouring hybridization. In other words, we can postulate that:

(a) human cultural multiplicity does not descend from lack but rather from phylogenetic redundancy, that is, from a complex and articulated ethographic structure;
(b) this redundancy does not increase the human delimitation to a present range of ontogenetic outcomes; on the contrary, it allows for a very wide space of evolutionary virtuality;
(c) some factors characterising the human phylogenetic structure prepare man to introject external references and thus transform virtuality in a disposition to malleability.

Thus, we can subvert the role of phylogenetic content, which is no longer deterministically connected to an ontogenetic outcome, but can be interpreted as openness to unpredictable and manifold evolutionary ways. This leads us to the question whether cultural evolution is an ontogenetic process (the simple species-specific development of phylogenetic inherence), or, conversely, if we are facing a particular event that cannot be subsumed under ontogenesis. In fact, even if we suppose that the multiplicity of identity is ascribable to redundancy and not to a deficit, we still need to understand if redundancy alone is enough to comprehend the cultural variety of the human being. In other words, we still do not know whether phylogenetic poignancy can be read in a sociobiological light as a potentiality of human predicates—which would lead us to consider culture as unmediated evolution of species-specific heritage—or whether phylogenetic content can be interpreted as factors favouring human construction without determining its connotations. In this last case, we should not only speak of ontogenesis, but also of ontopoiesis.

In order to understand this, we should consider the evolutionary process that starts from phylogenetic heritage and leads to individual identity. As we already know, ontogenesis especially characterizes animals with parental care, which represents not only a performative content—looking after, protecting, feeding etc.—but also educational-referential content, such as:

(a) acting as models of behavioural profile;
(b) determining an evolutionary differential of phylogenetic propensities and tendencies through direct interaction with parents and the community;
(c) favouring specific experiential centrifugations through reassuring and supportive presence.

In interspecific parental relationships, the cub organises its innate evolutionary endowments through the scheme offered by its parents, so we can state that a species' identity dimension—realised through ontogenesis—is not the simple translation of the genetic recipe, but the use of this information in an evolutionary process within a species-specific milieu. When such interspecific dialogue does not take place, we can observe a deficit of information in mammals and birds, namely, the subject's inability to reach a full expression of its identity.

The subject's need of an ontogenetic milieu gives the erroneous impression of its being somewhat absent-minded in terms of identity, not remembering something that is within it—hence the Platonic idea of education as recollection of an a priori. Actually, the evolutionary subject is neither a blank slate on which to engrave contents nor a parchment you can unroll to reveal innate content. Rather, it is a dialogic entity that builds its identity using innate characteristics as evolutionary material and the milieu as mould, or as the ground on which to plant its innate characteristics. The dialogue with the mother and more generally with conspecifics plays the main role in this process. The liability of innate information to be structured through context-based coordinates allows for the appearance of a second form of environmental adaptation—the first being natural selection—i.e. ontogenetic conformation, which can introject the peculiarities of the specific situation and thus optimize the individual's fitness to its environment.

However, the very openness of the system would run the risk to widen the gap between different individual identities within the same species, creating problems of belonging and communication, interaction with members of the same species, phatic recognition, courting, social relationships and so forth. Therefore, from an adaptive point of view, an ontogenesis that can vary freely is preferable, while from another point of view, the one concerning group belonging, ontogenesis should be confined into the area of single species. Social learning, whose most important aspect is the dialogue with the mother, serves exactly this purpose: it reduces individual difference as it constitutes a sort of training that confines ontogenesis to a precise range of variation. In this way, parental care allows for a large adaptive virtuality, while simultaneously ensuring compliance to precise species-specific canons.

By analysing these ontogenetic aspects we can understand that social learning keeps the ontogenetic process inside the species canon. However when we evaluate cultural outcomes, we can observe that they are not only the result of species-specific social learning, but, on the contrary, they tend to overcome the canon of species. For this reason there is no isonomy between cultural development and ontogenesis, though it is clear that the very ontogenetic potential allows for the evolution of a cultural tradition. The variability of these results contradicts the principle of species-specific belonging instead of confirming it, as is the case with maternal-social learning characterizing ontogenesis.

Cultural dimensioning limits individual variation, but it is never ascribable to—rather, it opposes—canons of species, shifting belonging to areas other than the species. We are facing an evolution of ontogenesis itself requiring external factors to act in a dimensional way (as in social-maternal learning), but also to transport the individual to configurations that do not belong to the concept of species. Therefore, we must posit a phenomenon that is not only ontogenetic, but also ontopoietic. In other words, this is a transformation of ontogenesis: an evolutionary event that is autonomous from the construction of the species canon.

This aspect is particularly relevant to us because leaving the species canon implies a human decentering—in a very similar way to Plessner's concept of eccentricity—from man's ethography. The difference lies in considering such eccentricity in self-determining terms, as Plessner does, or as the result of an extra-human decentering force. In other words, what we are trying to understand is whether phylogenetic predicates, taken in their evolutionary autonomy, can explain the decentered outcome or whether, as I believe, they are "dialogic forces" or inclinations to dialogue with heterospecific beings that build hybridational bridges with otherness. This dialogic inclination has allowed the human being to hybridize with the external world and give life to predicates that decenter it from it species canon. In order to better understand the difference between simple ontogenesis (individual expression within the species) and ontopoiesis (openness to cultural decentering) we need to reflect on the dialogic principle of ontogenesis, since individual development is in mammals always the outcome of a process of reflection.

Reflection as a dialogic event and not as a solipsistic phenomenon lies at the basis of identity in all animals that, in their evolution, have taken the path of parental care—not just in humans. Reflection makes us aware of our dialogic nature and, as every expression of the here-and-now, it subsumes the seriality of the dialogues and brings our connection to otherness to the surface. We are used to thinking that our choices are the result of algid rationality and that each of us is the only creator of her or his identity. In truth, we also act following our emotional structure, namely, the dialogic roots that connect us to the world—an endowment coming primarily from our dialogic relationship with our mother. Therefore, we need to abandon autarchic mythopoiesis—which considers man to be a completely self-referential individual, the hero of his existential journey—and acknowledge that every identity has a dialogic nature: that is, identity originates in dialogue and is then organised for dialogue. Identity's heteronomy has to be applied to every typology of identity emergence, the reality of which we must acknowledge, trying to avoid both reading its structure as an autopoietic and essential entity and annihilating its presence in the name of a horizontal nominalism based on individuality and the here-and-now. Reflection is, therefore, the place where identity can emerge through dialogue with a reflecting otherness.

The maternal-conspecific mirror has a centripetal action, which maintains individual identity inside the canon of its species, thus balancing the centrifugal action of context adaptation, which widens the gap between individuals. There is also a third kind of evolutionary event, which exists halfway through species-specific training and adaptive variation and which, through animal epiphany, paves the way to ontopoiesis. In the encounter with heterospecific otherness, the mechanism of identification is only partially activated, because of heteromorphism. However, the animal mirror is a reflecting entity that cannot be reduced to a mere stimulus. The human being recognizes itself in the animal, even though only partially, but this experience does not strengthen the canon of species; rather, it creates a change within the canon itself. Similarly to intraspecific reflection, the relationship with heterospecific beings, especially when it takes place on intersubjective and epiphanic coordinates, leads to social learning. Also, it has a centrifugal action, i.e. it leads to the subject's decentering from the species canon. This decentering takes place thanks to animal characteristics and the kind of reflection it prompts. First of all, we should wonder why the encounter with a heterospecific being has a reflecting action (which is very different from a mere stimulus). The main point here is the fact that we "recognize ourselves in heterospecific otherness."

The encounter with animal otherness gives rise to identification and sympathy—in the sense of feeling together—or to a reflective event, namely seeing ourselves in the heterospecific being. In my opinion, the reason for this is the great empathic capacity of human beings as well as their propensity to projection, which founds all their mimetic activities that can be noted from a tender age. However, what I have stated about zoothropic characteristics is also true for empathy: there are many paths to the openness to animal otherness, and to concentrate on only one of them would be misleading. An fascination exists between human beings and other species that can turn mere observation into rapture. This magnetism for the heterospecific being can be seen in cave paintings, the styles and rituals of traditional cultures, in children and, in a more or less encrypted—that is, more or less altered—way, in any human expression. The heterospecific being is a place for confrontation, interest, fear; it allows for dialogue with the world. We always find a personified presence in the animal, which is never really translated into a stimulus, but plays the role of a mask: the theriomorphic being is the archetype of every mask. As such, man sees the animal as a face, an encounter, a reflection.

Facing nonhuman animals is an adventure—horror, wonder, vertigo, excitement, enthusiasm—from which every form of narration begins, as the dramatization of theriomorphic masks. The staging of adventure requires the human being to wear the heterospecific mask, that is, not only to take on an animal appearance, but also to reproduce its movements and attitude, dramatising the process of animal-becoming. The animal "mask" is an overall disguise: sounds, postures, moves, body transformations, proxemics, choreographies, and behaviours. We can find the presence of the animal mask not only in shamanic art, but also in rock paintings representing theriomorphic entities or in scenes of zooerasty, in zoomorphic ornaments, ritual dances, or sounds that have nothing to do with human language. The animal mask clearly represents the concept of "non-foreign-otherness" expressed with heterospecific reflection and reveals the metamorphic process introduced by the encounter. The most important aspect is that in this encounter the human being sees the non-human animal neither as an entity in itself, nor as total otherness.

In other words, the encounter with the heterospecific being is not limited to the phenomenal event, but becomes a real epiphany. This is possible for two specific reasons:

1. the human being sees itself in the heterospecific being, which means it recognizes that they share a common ground and this sympathy leads it to project itself into otherness;

2. the reflecting entity, in this case the animal, by the virtue of its heteromorphism—obviously we are not referring only to its physical appearance, but to the overall positioning of the subject in the here-and-now—works as a distorting mirror, i.e. it gives back a human image modified by hybridization.

Due to the combination of these two factors, the human being not only sees a bird flying, but it recognizes flight; it does not only observe the ritual of the crowned crane, but it also feels in its own body the undulating movement of its choreography. The projection into the other as a neighbour prompts reflection, but the reflected image is immediately distorted by the animal's heteromorphism. The feeling that the human being experiences during this process is that of possession: it is invaded by new predicates created by the re-organization of the heterospecific predicate within the human body. The animal mirror distorting the reflected image is epiphanic in the sense that it inspires new perspectives or predicative possibilities, which are displayed by the reflected image. Hence the feeling of being possessed by an external entity (an idea or a divinity) able to show us the way to new existential possibilities. Therefore, animal biodiversity turns into a labyrinth of distorting mirrors where each of them shows a specific hybrid image, namely, a new ontogenetic path towards human evolutionary organization.

In this way the Epimethean heritage, that is, the various morpho-functional dispositions of nonhuman species, can nourish the human Promethean metamorphoses, i.e. the possibility of ontogenetic experience outside of species-specific boundaries. Heterospecific reflection marks the beginning of a tradition: a hybrid ontogenetic dimension next to a species-specific dimension, the former being introjected by the latter in what we call "culture". The human being is no longer included into its own species-dimension and therefore he can no longer be understood solely through ethologic analysis. The cultural dimension is not ascribable to the species-specific dimensioning characterising apprenticeship (i.e. social learning occurring during the evolutionary dialogue with the mother and in general with conspecific beings). This is demonstrated by the fact that culture, in its ontogenetic coordinates, does not belong to species but varies from people to people and can be subjected to fragmentation and deletion if the specific flow of tradition stops. While species-specific learning emerges spontaneously from the dialogue with conspecific beings, and respects ethographic expressive standards, culture requires a specific flow of traditions to be sustained, by the virtue of its unrelatedness to the species canon. However, we should not believe that the animal mirror's distorting effect is only attributable to heteromorphism, because the human being projects a mental state onto animal otherness, i.e. a mixture of emotional, motivational, representational structures that also modify the structure of the animal mirror.

The resulting theriomophised image can be enlarged in its totality or only in some of its predicates, following the emotional-motivational investment, or it can be modified according to representational structures. In theriomorphic reflection man encounters his own identity, which has been metamorphised by the recursive dialogue between his own mental state and heterospecific otherness. He can recognize himself as a hybrid identity: something different than before, an entity that is in progress, contaminated and projected beyond reflection. His reflected image is not a simple overlapping of bodies: rather, it is the integration of different states. Therefore, culture is not simply an imitation of nature, just as theriomorphism is not the simple overlapping between man and zoomorphism. The theriomorphic image reflected by the animal mirror is the diachronic outcome of a dialogue where the very projection modifies the animal form and is modified by it. Its distorting effect comes from heteromorphism, i.e. from the heterospecific being's Epimethean content, but also from wonder-appeal, namely, the projection of man towards the heterospecific being, or, in other words, the non-foreign contents that man can see in the animal.

Projection can take the shape of levity in the encounter with flying creatures, highlighting their aerodynamic and limpid forms, or, on the contrary, it can emphasize gravity in the encounter with mastodontic creatures, focusing on their heaviness, telluric qualities, roundness, opacity. The heterospecific being is thus transformed by projection, so that the human being can recognize himself in theriomorphism as an identity that has changed due to three factors:

(a) its body, or rather, its previous perspective—which is not necessarily its ontic dimension;
(b) its projection towards the non-human animal, with the specificities expressed in the encounter;
(c) heteromorphism, or Epimethean specificity, i.e. the specific predicates of that specific animal.

However, if we accept that every new identity stems from a process of reflection mediated by otherness, and as such it is also a process of hybridization, every research for purity in identity becomes fanciful. In this sense, the very word "identity" is stripped of its etymological meaning, as every assumption is also a reflection. The human being cannot categorize its own nature because the very creation of identity needs otherness. By accepting animal epiphany we also acknowledge that the human being can only see itself through the animal mirror, even though, paradoxically, what it sees is a theriomorphic entity, an over-human.

4.2 Identity as Hybridization

To understand the ontopoietic phenomenon—that is, the ontogenetic coordinates that do not enclose man into his species canon and give rise to a dynamic and unstable profile—we should not only consider the evolutionary dynamics, but also the processes of identity creation that work through the introjection of external factors during both phylogenesis and ontogenesis. The human relationship with external referents constitutes a represented identity that goes beyond mere somatic functionality. In other words, we cannot distinguish between man as an embodied entity and all the factors that have defined him in the process of somatization. The feeling that something is missing comes from a perspective error: that is, the presumption that the human body can be deprived of the external elements that co-constitute somatization.

Like other animals, man creates his own environment through somatization, therefore his body cannot be evaluated separately without feeling that something is incomplete. The error lies in considering the somatized elements as external to the body for the only reason that they do not organically belong to the body. This prejudice places the body in a pure and universal dimension, untouchable by external referents, and marks its difference from the strong performative declination of the non-human animal, which, on the contrary, is assimilated into its ecological milieu. Clearly the error consists in self-presentation, i.e. how man defines what is proper to him, through confrontation with, and reflection on, otherness. So we return to the paradigm of incompleteness, or the ideological manifesto of the human being's "feeling that something is missing" when facing animal performativity. All anthropological theories, varied as they may be, come from the same presupposition: the idea that the human being is disconnected from direct biological functions, namely, that it lacks correlation or gravitation into a precise range and reaches out to peaks that are precluded to animals, being free to act according to its will, while, simultaneously, being unstable and in need of protection.

Promethean mythopoiesis helps us comprehend this perception of identity that lies at the basis of humanistic thought and generates:

(a) the dichotomy that sees the human being as opposed to the animal;
(b) an instrumental and ergonomic vision of *techne* as an extension of the human predicate.

Myth presents an Epimethean legacy (virtues embodied by animals) through a collection of morpho-structures that give a rank to every species, while the human being is naked and owes all he has to the trickster Prometheus and his gifts, *techne* and fire. Human nakedness is the assertion of his *vacatio*, but only at a first glance: if we look more closely, it allows the *anthropos* to be seen as a protean entity that holds its fate firmly in its hands. The virtuality started by the apparent Epimethean amnesia makes man autopoietic, free in choosing his own forms, but also able to keep the performative declination outside himself, so as not to be contaminated by it.

The technosphere, then, allows not only for the vicarious function, but also for the separation of the human being from performativity, as it supports an undeclined entity through Promethean externalization. Performance is outsourced to the tool, freeing man from any kind of functional declination, like claws for preying or hoofs for running. The human being has no proximity to Epimethean children, who are relegated to the theriospherical biological dimension: while the theriosphere condemns the animal to its own performative category, the technosphere exempts man from having any performative category. This myth generates a neat disjunction between the human being (free, virtual and autopoietic: a key player in the world) and heterospecific animals, absorbed by the world as any other physical phenomenon and hence neither free—since they are predetermined by their performative declination—nor autopoietic (since they are unable to refuse their natural condition and empower themselves through their will-capacity).

To enable this identity operator, however, it is necessary that we reset our phylogenetic contents and give *techne* an ancillary role and a fully externalized dominion. In other words, the human being needs to keep an embryonal profile (far from adaptive specialization) and its Promethean artifices must be non-infiltrating and non-declinating entities, external and able to increase the distance between man and the world. Humanism is founded on a metric and subsumptive conception of man: the vitruvian anthropoplastic asking the human being to be neutral and play the role of universal measure, capable of taking on any shape. To do so, man has to avoid being contaminated by the tool, envisaging it not just as the means to a function, but also the glove that protects his hand from the function itself, keeping it pure and undeclined. In this case humanistic thinking is not just a way to conceive of man through separation from nature, but it also represents a specific way of considering *techne*, and more generally, culture.

To understand the humanistic mindset, we should take into account the interpretive model that sees nature and culture as two foreign and opposing entities, only connected by inverse proportionality. If we read the nature-culture dialectic from a dichotomic-complementary perspective, it becomes obvious that those who want to highlight the cultural character of the human being will tend to devalue the human phylogenetic heritage. On the contrary, those who believe in predicate heredity will not take into account evolutionary variables. This tendency represents a very serious error that has generated misleading readings of Darwinism. In fact, dialectical-evolutionary processes are based on a direct (not inverse) relationship, since the structure that presents a higher evolutionary virtuality—i.e. that can more easily internalise external influences—is the redundant structure.

If evolutionary adaptation can be considered the specific mirroring of the entity's growth conditions and not the simple filling of an empty container, then the more complex the phylogenetic evolutionary organization, the wider the horizon of reflection. Indeed, the most complex species from an ethografic point of view, like mammals for instance, are also those with the highest learning faculties. This should be enough to undermine the idea of inverse proportionality. From this we can infer that, contrarily to what is stated by the humanistic paradigm, the very phylogenetic complexity of the human being allows for:

(a) a wide range of possible ontogenetic outcomes;
(b) the magnification of the evolutionary influence of external social, contextual or heteronomous factors;
(c) the predisposition to the introjection of external references.

In its biographical and anthropopoietic nature, the human being is the result of evolutionary events that require a complex ability to introject external information. Supposing that our species' evolutionary specialization is referred to dialogical-mimetic capabilities and hybridizing tendencies, as well as to a great neurobiological redundancy allowing for many evolutionary paths, two facts become manifest: on the one hand, a substantial phylogenetic characterization and, on the other, the directly proportional necessity to turn to ontogenesis, namely, evolutionary processes, to understand how human complexity and variability develop. In contrast to philosophical anthropology, I believe that we cannot leave a phylogenetic assessment of human beings out of consideration—I will extensively speak of human nature—but this does not mean that we should read such complexity as a deterministic closure and an appeal to an inevitable biological reductionism.

Our species presents a compelling anatomical-functional complexity and a performative declination of its structure and physiology. How can we state that a neurobiological morphology entailing 100 billion neurons, which already presents migratory indications in the central nervous system and synaptic wiring specificities, is incomplete? How can this species be incomplete, if it has a practical/gnostic capability enabling it both to use a microtome and grab a ball that moves along a trajectory at over 150 km/h, with the same precision? Or if it has a vocal apparatus that can rigorously reproduce a high number of phonemes, through a complex structuring of its pharyngeal tract and a re-organization of its supralaryngeal apparatus? Or, again, if it has such a specialised visual apparatus that it allows for both reading and a binocular vision enabling it to perfectly calculate distance? What do we mean, then, by "incompleteness"? We cannot just state that, analysing man's morphological and behavioural structure, we can observe a remarkable level of specialization and complexity. If man feels he is lacking something and is distant from direct functional correlation, there must be a reason.

Therefore, we can identify this feeling through various somatization perspectives, i.e. through the introjection of external entities that, despite not being absorbed by the organism, *de facto* modify its somatic structure. I will here consider two coordinates:

1. the phylogenetic one, which hypothesizes that our species has undergone an evolution towards performative externalization, for instance through the use of tools. In this view, there are activities in which some of our organs are not suitable for direct action, but are fit for action mediated by tools;
2. the ontogenetic one, which hypothesizes the entrance of nonhuman others or tools into the evolutionary milieu of the human being: hence their influence on its evolutionary direction in the construction of its identity.

These two kinds of somatization show us that the human body is no pure and universal entity; on the contrary, it adapts to external factors. Taking point 1 into consideration, we must rethink the hypothesis of the reversive effect, but we must do so starting from Patrick Tort's reflection[2] on Darwin's evolution of prosocial behaviours in line with the fitness precept.

---

[2]P. Tort, *L'effet Darwin*.

What is interesting, even though not explicitly stated, in this line of argument? The idea that the subject's behaviour interferes with selection, modifying a species' morphogenetic canon. In other words, we are speaking of a pseudo-Lamarckian effect: the subject does not directly modify the species through inherited characteristics; rather, it does so indirectly by acting on selective pressure. However, the concept of reversive effect as screening presents many weak points. Natural selection acts on performative canons in their specific expression and context. Any expression—deriving from accident, mutation, sudden occurrence, invention, or something else—determines a performative canon and a consequent selective pressure. For this reason, rather than reversive effect we should speak of a modification in the performative canon: actions that used to be carried out in a given way are now replaced by others.

The performative canon can be introduced by culture as well as by genetic mutation or the epigenetic context: for instance, with the invention of antibiotics, selective pressure has moved from the body's ability to be effective in antibiosis to that of interfacing with, and tolerate, antibiotics. Every modification of the performative canon determines a consequent change of selective pressure—not its encryption. The performative canon expresses the what-to-do and how-to-do of adaptive processes; it can change the dynamics and areas of competition, the areas of externalization of functions, the counterparts we can adapt to and the contexts we can reflect. From this we can understand that every modification of the performative canon does not nullify the coordinates of selection; rather, it shifts them. This means that the selective pressure that has shifted away from a morpho-functional conformation acts upon another conformation, not creating primitivism, but increasing somatic specialization. In other words, the techno-mediated performative metamorphosis influences the "organ-instrument-function" triadic relationship, where the selective pressure has moved onto conforming both the organ to adapt to the medium and the latter two to the new performative canon.

For this reason we can affirm that every cultural intervention does not nullify selective pressure: it just transfers it, making species even more specialised as conform at a higher level gradient—the body adapts to the tool and the performative couple adapts to the function. But this is not the only point. In order to give life to a relevant hybridization, we must necessarily assume the existence of phylogenetic predicates creating strong dialogic-hybridizing inclinations. Therefore, although we cannot deduce the cultural outcome from the phylogenetic heritage, species-specific characteristics have promoted the very hybridizing process that has given life to culture. After this clarification, I believe the major effects on culture on identity should not be attributed to selection, i.e. to a direct intervention on heritage, but rather to the developmental coordinates (point 2) that translate innate organization into an identity profile. There are two coordinates of evolutionary modification:

(a) the epigenetic level, that is, the context of translation of genetic information, especially during embryogenesis but also during growth;
(b) the ontogenetic level, that is, the construction of the person's cognitive-behavioural identity.

There is no doubt that these coordinates come into play when man has a feeling of lacking something when being deprived of his hybridizing media. Such coordinates make him feel *ex post* the need of cultural heteronomy, i.e. of the contribution given by media to the expression of predicates. In other words, this sense of lack can be referred more to a retrospective perception than to an actual lack or incompleteness of the human nature, since we recognize ourselves in the human performative (i.e. ontogenetically hybrid) canon, and not in the phylogenetic one. With respect to coordinate (a), it is increasingly evident that the phenotype is not the simple translation of genetic information, but rather a particular way of translating it. For instance, height depends on the diet, so that times when the population is taller and others when it's shorter succeed one another cyclically and with a sinusoidal pattern in human history, due to the availability of certain resources. At the same time, a development contraction of a mother has effects on the size of her pelvis and this, also, influences the size of the foetus. Diet and the mother's bodily structure have epigenetic effects and we know that cultural and technological transformations have revolutionary effects on eating habits.

However, agro-zootechnical modalities can have epigenetic effects if, in that particular plantation or farm, chemicals like auxinics, pesticides or hormones are used. Also, an environment modified by technology can determine exposure to electromagnetic waves, radiation and other factors influencing epigenesis more than mutagenesis. There are customs like bandages, bustiers, specific ritual gymnastics, the use of particular objects inserted into the body—neck rings, lip plates, braces—that modify not just the appearance but also the epigenetic context of phenotypical translation. Even the synaptic and immunological structures are modified by variations in the epigenetic context, to the point that we can state that today's "digital natives", who live in almost aseptic environments and undergo very specific stimulations, have completely different neurobiological and immune structures from teenagers fifty years ago. To briefly explain culture's influence on the epigenetic context is almost impossible. What is certain is that morpho-functional conformation is an epigenetic product, and not just in the conclamation of teratomorphies, as is the case with phocomelia coming from thalidomide usage. Culture is a co-factor in the formation of the amnios in morpho-functional development, and this is central to self-perception. Indeed, we recognize ourselves because we are modified by epigenetic influence, which is determined by cultural dimensioning.

Coordinate (b) regards ontogenesis, i.e. the construction of the person's identity, which is linked to parental relationships, pedagogical coordinates, social connections and experiences that are enabled or fostered by a certain culture. Cultural dimensioning and parental care determine evolutionary coordinates, i.e. the spaces for development and, thus, the ontogenetic outcomes. Therefore, in the cultural partner we can identify the same operators that work in the maternal relationship: (i) the "secure base" effect, or the experiential facilitation achieved in some selected contexts; (ii) the proximity effect, or the identification of an evolutionary space that can be covered during the evolutionary process; (iii) the referential effect, namely the production of models influencing identity construction; (iv) the expressive field effect, or the emergence of precise spaces of individual expression, which influence both self-perception and identity evolution. Ontogenesis, as the overall construction of individual identity and its perception as a biographical entity, is influenced by the cultural milieu starting from an early neonatal age. For all intents and purposes, ontogenesis is a construction and a recognition of the self, realized through mainly cultural coordinates.

Every culture not only defines group models and canons: it also establishes evolutionary directions and self-recognition structures. Just as we cannot eradicate the subject from the dialogue with its mother, similarly we cannot consider culture as a superstructure or a set of customs and traditions. The individual's identity structure has developed within a precise cultural milieu, and thus it has adapted to it: the subject recognizes itself through that particular dimensioning. To separate the individual's identity from its cultural milieu would be like looking for the heart of an onion by peeling it. To use another metaphor, the dress has become a part of the body, just like the skin: it concurs to the creation of identity. If the subject is deprived of it, it will not recognize itself, its identity. Of course, in such a perspective, the very idea of removing the cultural apparatus makes the subject feel naked. Self-perception, in fact, can be traced back to cultural dimensioning, which introjects the cultural morpheme, so that the subject is alone in that particular morphology. To speak of culture as a superstructure means not understanding the workings of the ontogenetic process—that is, the construction of individual identity through the introjection and reflection of experiential conditions.

To hybridize with external referents means to inaugurate "ontopoietic dimensions", i.e. spaces of development of ontogenetic coordinates, allowing ontogenesis—as the construction of individual identity—to escape the static nature of the species canon and become an evolutionary entity. Ontopoiesis is, therefore, a sort of meta-ontogenesis: a metamorphosis of ontogenetic coordinates created through the insertion of evolutionary seeds in social learning. In this way, dimensionality is no longer a definite and stable space—the species canon—but an evolving universe, unstable and metamorphic. Ontopoiesis is the continuous formation of possible ontogenetic coordinates that can dismantle the canon of species and the connection between the individual and its ethogram. On the one hand, this transformation of the ontogenetic space—from a delimited reality deriving from phylogenetic heritage to an infinite space, free from any boundary, and especially no longer inferable from the phylogenetic heritage—reduces species-specific gravitation, while inevitably creating the need for new identity spaces. This new ontopoietic emergence allows the human being to explore new existential dimensions, though always maintaining the connection to its phylogenetic heritage. Deviation from a shared species canon creates the need to cling to a new—cultural—identity. In other words, heteronomy becomes the cornerstone around which the expression of identity gravitates, changing the human profile from a species-based ontology to one centred on the relationship with non-human beings.

4.3 Hybridization, or Falling in Love

After these necessary clarifications, I must say I believe the feeling of lacking something is the simple effect of perception that can be connected to the presence of cultural mediators, which modify the standards of perceived optimality, so that the very idea of depriving ourselves of these mediators would lead us to performative efficiency-effectiveness standards that are below our expectations. The emergence of a new dimension in the formation of human identity thanks to theriomorphic projection or to new technology is similar to the feeling of falling in love: missing the partner, clearly, does not precede infatuation, but follows it. This lack is always an *ex-post*—never *ex-ante*—situation; therefore, the Promethean reading can also be ascribed to the teleological bias. Depriving ourselves of the hybrid dimension means falling into a condition of non-recognition, not because the external referent would bring us to a state of primitive nakedness, but rather because our existential dimension has absorbed otherness, conforming to its morphology.

To better understand the feeling of lack, I will make reference to two coordinates of predicate attribution:

1. the subject's predicative dimension, or the very perimeter through which the subject evaluates its own presence and, consequently, the feeling of missing its techno-mediated dimension;
2. the definition of the objectives or goals set by the subject and the recognizability it attributes to the techno-mediated image, i.e. to what extent the techno-mediated canon has been introjected during the process of self-recognition. The feeling of lack can be caused by simple routine: you wouldn't need to be a "digital native" to perceive lack if, suddenly, you were deprived of all the IT devices through which we have built our life dimension. This feeling cannot be described as an a priori situation (preceding cultural emergence) since, from a performative point of view, it is the very producer of the lack. Only after the invention and diffusion of mobile phones have we reached the performative condition that makes us feel we would miss them if we no longer had them. For these reasons, we can state that culture does not fill a gap: rather, it creates it.

Elsewhere I have examined how cultural emergence creates a real addiction, leading man to non-identification with himself.[3] Therefore, the mistake would be to place the phenomenon a priori (the exemption from an original lack) rather than a posteriori (the emergence of a need due to the new performative canon). Therefore, we can state that culture is not the remedy to human incompleteness or imperfection; rather, it is similar to falling in love, in the construction of a hybrid space for expression connecting us to otherness and making us miss our partner. With regards to coordinate (1), cultural emergence operates a real slip in human ontological perception, since it:

[3]R. Marchesini, *Il tramonto dell'uomo. La prospettiva postumanista*, Dedalo, Bari, 2009.

(a) creates new predicates that did not exist in the species canon;
(b) establishes effectiveness-efficiency standards in the modalities of self-awareness and awareness of the world;
(c) defines the collection of assumptions of social recognition;
(d) plays the role of judgmental element in social relationships.

These four factors suggest that there is a deep connection between cultural emergence and the subject's operative ability, and especially with its self-perception. The cultural milieu has a psychedelic action, in the sense of "expansion of consciousness",[4] similarly to the way in which somatization represents an expansion of the body. The human being is present as a subject not because it is inherently able to distance itself—which would imply that every cultural expression and especially *techne* are only consequences of the human being—but because it can expand its consciousness beyond itself through the action of hybridising factors. Thus, culture enters the human predicative milieu like a virus, shifting the limits of his consciousness.

(1a) The subject's presence in the world—its *Umwelt*—is connected to immersion, or the perceptive modality of external reality. Consider some dimensions of Dasein that are modified by cultural emergence: to observe a virus through an electron microscope; to go to the moon on a spaceship; to reproduce the image of an event with a video camera; to keep in touch simultaneously with hundreds of people from every corner of the world thanks to the World-wide Web; to send pictures and documents, transforming them into packets of information thanks to computer and phones; to transfer a gene from the human genome to a bacterium in order to produce insulin through biotechnology. These are all blatant examples of predicative dimensions that the human being could never reach without the cultural partner. However, these considerations are also valid for the great part of post-Neolithic creations—such as diverting a river, cultivating a field, building a house, creating pottery or weapons, or hunting. If we were deprived of culture, we would immediately lose some interfacing dimensions that we consider integral parts of our human condition. To cancel a certain cultural dimension means to fall into a predicative regime that we would consider subhuman and inhuman exactly because it lacks those performative characters that today we perceive as typically human.

(1b) The various cultural apparatuses establish the level of effectiveness-efficiency I should expect from my projection into the world: for instance, after the invention and diffusion of mobile phones I find it is normal—and therefore I demand—that I can reach anyone anywhere at any moment. If I forget my mobile phone or the other party cannot be reached, I have an immediate feeling of *ineffectiveness of presence*, and only later *of performance*. Similarly, if in order to contact someone I have to send her a letter or find a phone box, I have a

[4]Cf. H. Osmond, "A Review of the Clinical Effects of Psychotomimetic Agents", in *Annals of the New York Academy of Sciences*, Vol. 66 (3), 1957, pp. 418-434.

feeling of performative inefficiency. This means that the subject perceives itself not only in terms of predicative catalogue, but also in terms of gradients of presence, which are highly influenced by cultural mediators. If we only consider the performative level—as the capacity to act in the world—we will realize to what extent this parameter is influenced on a perceptive level, as well as objectively, by the cultural factor. The performative level shows the type of result and how to reach it; the shift in performative gradient operated by cultural emergence makes us inevitably feel meagre whenever we imagine not having technological devices. In other words, cultural emergence determines a shift in predicative perception in the broad sense; it modifies the way in which man recognizes himself and defines new parameters of belonging. Besides, every expression is measured and standardizes on the basis of socially determined levels of performativity and judged on the basis of culturally mediated validation. Cultural mediation defines the perceived standard of presence.

(1c) The various cultural apparatuses of individual immersion define the collection of acceptance and the predicates of social recognizability. The following is an example of performative model: if a person does not have a car nor uses other technological means of transportation, she cannot take part in many social dynamics, from the most ordinary ones, such as having a job and a family, to the most casual ones, like going to conferences, concerts, seeing friends. The person who suddenly stops using technological means of transportation will face problems of effectiveness (she will have to stop certain activities) and efficiency (others activities will become more difficult). In such a case, not only would the individual lose a range of performances and gradients, but she would also miss the opportunity to take part in the social forum, finding herself in a condition of social exclusion or deprivation. However, this analysis, which is based on the performative meaning of the cultural apparatus, despite being a good example of the concept of active participation in social life, does not explore the significance of belonging that concerns the overall individual rather than the simple performative aspect. A person is accepted on the basis of expressive coordinates regarding, for instance: the body as milieu and point of convergence of lines of social interaction, her involvement in rituals that determine membership, the definition of genealogical canons, or the adoption of habits and styles.

(1d) Cultural apparatuses do not only establish predicate collections, presence levels and social acceptance features, but they also become organs capable of evaluating the subject's predicative canon: for instance, today the people who do not use computers are cut out from the most part of work and some social activities. In other words, computers are the judging tool of the individual's predicative ability. Going back to the simple performative example, we can state that technology defines the parameters of somatic congruity, i.e. what body characteristics are preferable or optimal for a certain techno-mediated performance. In other words, as much as ergonomics has tried to create a technology that would adapt to the body, we can see that the opposite effect has occurred: the body has adapted to technology. If it is true that this has always been the relationship between the body and external referents, today this acknowledgment is even more obvious as there are technologies requiring the body to have a coherent and welcoming interface. Therefore, the predicative canon introduced by technology defines choice parameters and rewards the most adequate conformation.

As regards coordinate (2), there is no doubt that the human being recognizes itself in an identity project, which presents factors influencing and correcting the route of identity through operators such as:

(a) the need to highlight our characterization and broaden the distance between ourselves and otherness, favouring disjunctive elements;
(b) the tendency of the later steps to be influenced by the previous ones, namely, the self-implementing and segregating effect of choices and culture's autocatalytic effect;
(c) the non-equilibrium determined by decentering, which makes identity increasingly unstable in the internal cohesion of predicates, chaotic in its trajectories (and therefore susceptible to singularity effects), as well as more vulnerable to external intervention and increasingly needing heteronomies;
(d) the emergence of new hybrid identity goals, where the cultural partner has no longer the role of facilitator of human predicates, but rather introduces new objectives in the development of human identity.

(2a) Undoubtedly, the definition of man as a deficient entity, as *minus habens* compared to other species' performative gigantism, is coherent and functional to the construction of the heroic myth of the human being. Despite starting from a disadvantaged condition and having to deal with a highly superior counterpart, man has unexpectedly been able to win the challenge. Lack or deficiency is, therefore, an important side to anthropocentric mythopoiesis. Moreover, to consign the human being to a Promethean dimension, thus differencing it from all other species—confined to a rigid Epimethean declination—means to strengthen the identity disjunction. Lack becomes the *limes* separating the non-human barbarian world from the human cosmopolis, the operator allowing the human being to construct entire worlds, have control over its drives, and emancipate itself from the telluric dimension. To consider culture as a performative compensation, foreign to non-human influence, defining the human *telos* as an impervious entity, is much more productive than admitting hybridization. Humanistic thought is based on disjunction as a process of human cleansing and enucleation; hence, any admission of a contaminating action by external referents would bring inconsistency and contradiction to its very message. However, the more theriomorphized the human being, the broader the distance from its "nature"—in the sense of *Homo sapiens*' phylogenetic heritage. Such a difference, which is not a dismissal of nature *in toto* but only of our species-specific dimension, is interpreted as a generic separation from the animal condition, which, on the contrary, is taken to be totally immersed into nature. We tend to confuse the two terms (human nature and general nature) because of the humanistic dichotomy between natural and artificial, which misunderstands the very concept of nature. However, in this way we operate a conceptual distortion, since we attribute the coordinate of divergence with a heightening and emancipating directionality. On the contrary, if we accept the principle of animal epiphany, we will also acknowledge that, when departing from its own nature, man gets closer to nature in a general sense, establishing a tighter connection to other species. Indeed, introjecting external references, man mirrors nature even more.

(2b) Traditionally, culture is taken to offer a "crutch" to human ontic, i.e. it is responsible for the implementation of balanced situations, however, the truth is that it has the opposite effect. Culture is autocatalytic, since with the creation of non-balanced situations and the introduction of a magmatic element within the human system, it paves the way to new cultural expressions. If we dismiss the humanistic conception, which considers *techne* as being at the service of inherently human purposes, we can observe that, like any other culture, technoscience modifies human ontology. If *techne* shifts the interface dimension, it also produces a psychedelic effect. This catalysis takes place in its own ways that cannot be controlled or planned by man. The more we proceed with technological development, the further away will this catalysis be from human control. This loss of control—already perceived in the Nineteenth century, but overtly proclaimed only during the Twentieth century—certainly contributes to the magnification of the perception of lack. This situation is unavoidable, and the thought of an escape only produces a feeling of shortcoming and ontopoietic nakedness, since it seems we are escaping our own ontological fate. Culture appears as a sort of virus using man as replicative environment, shaping his morphopoietic route, where the increase of media makes the infection worse. This metaphor, despite being used by many authors, may seem far-fetched; however the feeling of lack is undoubtedly connected to something transcending performative inadequacy and regarding the immersion into *techne*: it is the feeling of not holding the helm of human ontopoietic navigation.

(2c) This feeling can be also ascribed to the non-equilibrium created by cultural dimensioning. Decentering, in fact, increases heteronomy as it shifts the focus away from phylogenetic heritage and especially from the canon of species. The theriomorphised human being is not just unstable and more apt to rely on nonhuman references: it is also more sensitive to further contributions from the outside. The psychedelic action of animal epiphany makes it increasingly more open to non-human conformative action. This is what I call "zoothropic circle", namely, an intensified orientation towards the heterospecific referent after a zoomimetic process. To leave the species-specific gravitational centre means to reduce the forces keeping the human being into a limited range of ontopoietic possibilities. This means that the human being perceives itself as an increasingly magmatic entity, eradicated from a specific phylogenetic heritage and available to operations of somatization and psychedelia.

However, we should be aware that such plasticity is not indicative of an increase in human autonomy—the heightening emancipation *à la* Pico also dominating Slotedrdijk's theory—but rather suggests its opposite.[5] The epigones of late Humanism interpret the apparent lack of roots and ontopoietic proliferation as shortage, neutrality or deficiency of a particular positioning. In other words, the feeling of lack is connected to ontopoietic proliferation, but mistakenly so, since organizational plurality, i.e. the possibility of taking on different forms, always results from releasing actions over phylogenetic structures. If a certain motivation is separated from its biological function—for instance, the syllegic propensity from its expression in food collection—we will immediately notice new forms of ontopoietic declination. Animal epiphany, with its decentering action, operates a separation and marks the beginning of an extremely magmatic situation that facilitates new epiphanic processes.

(2d) Undoubtedly, the rise of technology not only has a strengthening action on human phylogenetic predicates, but also allows for the emergence of new predicates. The theriomorphic or technological hybridizer inaugurates overhuman predicates based on the hybridization with other species: music, dance, cosmetics, fashion. These are all examples of animal-becoming, i.e. taking on non-human predicates that turn the ontogenetic dimension into a trans-specific space. In this sense, if we think that the word "man" refers to a specific phylogenetic entity, then clearly the cultural process determines the birth of an over-human entity, that has nothing to do with Nietzsche's *Übermensch* nor with transhumanistic transcendence. However, when we internalize overhuman predicates, we inevitably modify the human structure itself. In fact, the overhuman predicate has a marginalizing effect (establishing a domain of validity) on the species perspective, which is confined into a specific dimension, becoming partial if not erroneous. The human being is thus projected beyond the human, tracing new coordinates which make it obsolete not because it has been elevated to a spiritual dimension, but because it is increasingly more theriomorphic.

[5]P. Sloterdijk, *You Must Change Your Life*, Polity Press 2012.

Chapter 5
Zoomimesis: Embodied Epiphany

There is a particular way to think of ourselves as human beings, which consists in denying our phylogenetic condition and sublimating our being inevitably circumscribed within a taxonomic perimeter. The human being must counterfeit itself in order to feel its humanity: it has to modify its skin, change some of its anatomic details, gain a kinaesthetic sense that does not belong to it, transfiguring survival strategies and altering the way it uses its voice. Anthropopoiesis, as a kind of metamorphosis that takes the human being outside its species-specific shell, is an act of denial of our biological condition rather than an attempt at compensation. In fact, what kind of compensation can there be in music, dance and cosmetics? None, of course. There is only the need to transcend our dimension. Being human means dreaming to be elsewhere, distancing ourselves from our nature. This, different from what Plessner thought, as it does not originate from Plessner's doubling, but from being reflected on a therianthropic level: only an epiphanic projection onto otherness can allow for reflection. Decentralization is like passively letting the other take us by the hand and lead us, which is very different from firmly holding the reins of our own path.

Moreover, since in cultural emergence there is neither internal necessity nor determinism, ethnographic studies based on exclusively naturalistic premises cannot understand the real meaning of anthropopoiesis, that is, the indefinability of the outcome. I have highlighted some possible causes of anthropogenesis in the dispositional structure and cognitive characterization of the *Homo sapiens* species; therefore it would be incorrect if we did not consider the phylogenetic characteristics of our species in order to understand anthropopoiesis. The redundancy of our legacy and the tendency to be conjugated and internalize external references are what promotes the epiphanic event. However, we should also distinguish such inevitability (more than simple probability) from a state of internal necessity (fulfilled *motu proprio*), which would erroneously lead us to explain the anthropopoictic phenomenon separately from the intervention of external influences. When I state that there is no internal necessity in this process, I mean that we cannot explain the anthropopoietic event, both in its taking place and in its cultural

R. Marchesini, *Over the Human*, Numanities - Arts and Humanities in Progress 4, DOI 10.1007/978-3-319-62581-2_5

outcomes, by only relating to anthropogenetic factors or any other cause that is internal to the human being. It would be incorrect to ground the human being *iuxta propria principia*.

Anthropogenetic factors like the tendency to mimesis or a good neurobiological-evolutionary plasticity give life to unavoidable processes, such as: orientation towards heterospecificity, feeling together with animals, recognizing that we are united by a common animal-being, wonder for pteriomorphia, projection into the animal body or possession of one's own body by animal predicates, epiphany and the envisioning of new existential dimensions, as well as the introjection of non-human animal models. However, these processes are strongly related to external intervention. That is, they do not respond to internal necessity, unlike, for instance, the fall of an apple or the mechanic expression of an algorithm. Rather, they respond to a dialogic predisposition that has no autonomous evolution, but requires the intervention of an external element, without which the process would be unintelligible. After analysing the anthropogenetic factors and calling the autarchic theories on the anthropopoietic process into question, we need to understand the other source of human emergence, i.e. the dialogic otherness that favours the processes of anthropopoietic metamorphosis.

Going beyond man and adulterating the phylogenetic content with non-human contents necessarily requires that these allochthonous elements have the ability to be integrated into the human canon, which means they are not so alien as they are usually taken to be. Indeed, a given character must be recognized (that is, correctly attributed) before it can be applied in overlapping to the human ontological dimension. Painting one's skin, wearing a leather or feathered ornament or following theriomorphic choreographies, as well as using a predicative style that is typical of heterospecific beings, are all processes that imply both a consonance—recognition and body representation—and a dissonance, i.e. the perception of unexpected, foreign and challenging elements that call for introjection. Animal epiphany is a recursive process of assimilation of difference:

(a) seeing oneself in the non-human animal through a metapredication of commonality that brackets the predicates of difference;
(b) feeling excited for the alienating effects of heteromorphic projection or heteronomic possession;
(c) theriomorphic somatization or psychedelization.[1]

Briefly, in order to enter the non-human body we need to start from a communality, i.e. a consonance within difference: we need to go beyond heteromorphism

[1]An example is dance. In order to understand-feel a non-human animal choreography, marked by precise rhythms and specific movements, you need to find a harmony with the animal observed. Also, (a) dissonance is implied in repeating the choreography through the imagination or by the immediate expression of the body (b); in order to translate dance from the functional context of the species (the expression of courtship) as a non-human predicative dimension—other-than-oneself—into the representation of a possessed self (other-with-oneself) an operation is required to reduce the introjective dissonance (c).

and feel a common ground beneath the surface. To feel stupefied at the dissonance and the expressions of difference in our own body produces alienation, so we need to seize this therianthropy and reduce ontopoietic dissonance. In epiphany, the heterospecific being is not other-than-oneself, but other-with-oneself; an otherness—in Lévinas' words[2]—that mirrors the person (albeit reflecting an altered profile) and generates a hybrid image that still shows a good level of recognisability. In the forms of angels or sirens man does not lose his identity; in somatization and psychedelia he does not risk disintegration, but finds a new existential dimension.

What interests me is what is proper of human beings, which is neglected by the whole myth of compensation in the mistaken belief that only a frightful biological void could hold such a large cultural content, or that only a large inability and exposure to the world could demand-allow such a detailed and accurate work of exemption. On the contrary, I believe that human ontic redundancy and its wide ontogenetic range allow for this meta-ontogenetic opening—namely, the development of the evolutionary possibilities of human identity. We move from a complementative vision (summarized in the idea that only what is deficient can escape determinism) to a dimensional reading in which complexity is what allows for this redundant situation that, through evolutionary virtuality, avoids determinism.

Another issue requiring an explanatory inversion has been the humanistic precept by which identity has to be pure: that is, being human is the cleansing of and disjunction from non-human beings. Emanative and autarchic theories—or the humanistic paradigm's explanatory model—assume that we can explain the human being by referring only to man and his predicates; however, in a dialogic view we cannot enucleate man from its relationship with non-human beings. Enucleating man is not possible, not even if you consider man as being a completely spiritual entity or if you concentrate on the phylogenetic predicates of the *Homo sapiens*, in order to either show their vacuity or celebrate their meaningfulness. It isn't possible: not even if you suppose there is a self-founding process, an internal and self-referential necessity to mankind. It is not possible because any identity that arises from dialogue always requires a double recognition. If the human being is the hybrid outcome of man and non-human otherness, we must necessarily understand those human qualities that favour the dialogic disposition and introjective capacity, but it is also necessary that we understand how animal otherness can act as an interlocutor.

Undoubtedly, the flight of a bird has something majestic and impressive about it. And yet, if an eagle's large circles, an albatross' gliding, a hawk's diving, a hummingbird's mid-air stillness were nothing more than a phenomenon to the human being—just like a waterfall, a mountain, or a hurricane—as sublime as they can be, these events would not be able to announce a different existential dimension. For a bird's flight to be an epiphanic event, the human being must see itself in the flight: there must be an overlapping between the human being and the bird—the

[2]E. Lévinas, *Totality and Infinity*.

emergence of a bird-shaped man, or a reflection of the human in the bird. The flight enters the human body, which suddenly feels the convulsive movement of the bird's wings and the freedom coming from being able to fly. At the same time, man is projected into the bird's body and from that height he can consider with greater objectivity the narrow position of his own body, but also the vertigo caused by his new existential dimension. He feels the need to redefine his *Umwelt*.

Only through an aesthetic operation can the human being see itself, since seeing oneself means transcending oneself. However, in order for this overlapping to take place, there must be a correspondence between the human body and the non-human body, which means that the predicative differences—for instance, the bird's wings—do not prevent the correlation. To realize the epiphanic event, man has to find a metapredicative commonality that allows him to overcome the differences with the non-human animal and turn them into accessory conjugations of his own ontic structure. So, we can read the myth of Epimetheus and Prometheus in an opposite way to humanistic tradition: if the latter gave preference to a disjunctive interpretation, in my reading I would like to highlight the founding aspect of the former myth, that can be referred to the arising not so much of human beings as a species, but rather as the overcoming of man. In other words, the human being arises when man's dialogic and introjective tendency, of Promethean nature, meets the Epimethean epiphany, namely a non-human feature transposed into the human body. Hence, what is Epimethean is not so much the non-human feature per se, but rather the ensuing revelation, implied in the non-human predicative difference that comes across as an opportunity.

5.1 Recognizing Oneself in Otherness

The heterospecific being comes to us full of predicates of difference—morphological forms, chromatic and tonal variations, kinetic biomechanics, postures, styles, habitats and friendships, communication and interactions. In other words, it is characterised by other modalities of being-existing, different from ours and very adaptive, functional, peculiar, specific, and distant from our own. However, in these predicates we recognize—even before getting to know them—something familiar, as such predicates are supported by meta-predicates of similarity and sharing. Beyond the anatomical, physiological, ecological or expressive details, common roots arise, which are meta-predicative because:

(a) they lie beyond predicates;
(b) they found predicates.

Animal-being implies some very strong sharing meta-predicates, such as the experience of suffering, moving in search for something, interpreting the here-and-now, self-expression, vulnerability to the world, reproduction—just to mention some. By the virtue of these meta-predicative structures, the other animals, as different as they may be, are never totally foreign to us: we do not perceive them as alien, but rather as neighbours. When we encounter them, we are inclined to feel a certain connection to them. The fact that this connection does not necessarily imply respect is well known, and it should not come as a surprise. In fact, even towards our human neighbours, respect is not always the expected outcome of the awareness of similarity.

Sympathy—understood as the recognition of another's emotional state—is not enough, nor is empathy (understood as accepting another's different but equally shareable state) if we want to promote the following feelings:

(a) compassion, or "feeling together";
(b) altruistic behaviour as a motivational pre-reflexive disposition to help;
(c) the real deontological choice, understood as decision, duty, supererogation, often in conflict with our emotional or motivational disposition.

Therefore, what I would like to emphasize here is not the dispositional or even ethical meaning of this recognition, which is denied by the problematic relationship between facts and values, but something different, namely that the identification with animal otherness is not attributable to anthropomorphic projection—as is usually maintained—but rather to an effective meta-predicative sharing that the human being feels immediately, as indeed do other animals (although, perhaps, our species' great capacity for empathy strengthens this identification).

Recognizing ourselves in meta-predicates means going beyond the specificity of the act and recognizing the foundations of our belonging to the ontic definition of animal-being. While plants create their perfect bio-dialogic harmony with the sun, through the autotrophic power of photosynthesis, animals do not have any such privileged relation, so they are left to act as beggars in the ecosystem. To give an example, recognizing ourselves in the food meta-predicate means that, even if we are different in terms of heterotrophy (we don't rummage through leaves to pick fruit or berries, graze the grass, dig the ground for tubers, wait in ambush and chase the prey, etc.) we recognize a common ground: the need for food, the pleasure arising from eating it, the difficulty and risks involved in procuring it. Animal-being means grounding our existence on openness, in the awareness of a heterotrophy that makes us inevitably dependent on external biological mechanisms. It primarily means accepting that we are unable to survive autonomously and are destined to look outside for sustenance. In this inevitable struggle, we need to be able to recognize the competitive or predatory risk constituted by the other: recognizing each other is consubstantial to animal-being.

Autotrophs just have to wait, as their dialogue with the sun, albeit interrupted by pauses, is assured by celestial mechanisms; on the contrary, heterotrophs need to have appetitive and aversive sensors (i.e. approaching and distancing), as sensitivity and movement are the same thing. Suffering would be pointless if we couldn't get away from the source of pain, and there would be no need to move if we couldn't feel pleasure or pain. Feeling is the other side of movement: in this sense, every vain discussion on animal sentience becomes an idle argument. Feeling and moving, in their apparent dualism, are meta-predicates of animal-being, spaces where we inevitably recognize ourselves. In doing so we can easily transfer their predicates of specificity onto us: a child waves his arms pretending to fly or squats to jump like a frog. And thus, if flying, crawling, jumping, swimming, and running are incomparably different, still they share a common root that lies at the basis of animal-being: namely, being able to move to reach the sources of pleasure and get away from risks and sources of suffering, and, especially, the satisfying exuberance of movement itself, as the powerful feeling of taking possession of space in the magnificence of one's own presence.

Similarly, we cannot ignore the difference between the species' various *Umwelten*, since all accesses are defined by particular sensory windows, dispositions, perceptive Gestalts, or discriminative, cognitive, categorial and correlative functions, which are the outcome of particular phylogenetic processes. The species-specific context is the partial reconstruction of a world, obtained by accentuating certain feedbacks and blocking others, but always starting from objective reality, whose presence is demonstrated by the mere fact that it selected perception itself. Animal-being means sharing a world and living together in a reality that has conditions and points of resistance, which are transformed in selective pressures according to phylogenetic orientations. However, it mostly means sharing the reasons that lead us to reconstruct the here-and-now. For this reason we can state that, even though every animal lives in its own world—but not in a monadic way—in the reconstruction of the surrounding reality we recognize a common need of elective monitoring. If elections—in which fully expressing one's life means bringing out a useful-appropriate world made up of risks and opportunities—are specific, in the end we recognize ourselves both in the need to monitor and in the aesthetic pleasure arising from the confirmation of our perceptual expectations.

Biodiversity predicates are therefore areas of meta-predicative variation recognizable despite their connotative variability, which sometimes is very paradoxical, because the common root of meta-predicates—that is, what is declined in each species in a different way, but does not present any variation as such—is shared. Hiding in coral reefs, among the leaves of a tropical forest, in the den of a dark cave, in the depths of an ocean or in underground tunnels are simply different ways of doing the same thing. Animal-being means presenting the vulnerability of a thermodynamically unstable precarious structure that sways like a tightrope walker and is on the brink of falling into *post mortem* disintegration. Animal life seems to be a miracle and its persistence against entropy's dark dominion is assigned to the individual, which has to defend, with all its might, a miracle it has not asked for. Hence life's sense of strength and weakness, the gripping power that inspires any animal in its fight against the overwhelming, its everyday heroism, knowing that it can only delay the mandatory victory of death. Yet life has one last chance: reproduction, casting a living being beyond one's own individual space.

Here's another moving and striking meta-predicate: namely, the animal reproductive frenzy, the larval swarm of life that fills every space, luxuriant and restless, in a thousand forms of courtship, with its dances and colours. Reproduction shows us a life that wants to "show itself": it is aesthetics of the figure, organized in order to stand out from the background. Life is a phatic evidence in its desire for sexual recognition; it is exhibition and full self-expression. Reproduction is opposed to survival, which amounts to individual life, hiding one's vulnerability, life at all costs (which inevitably requires mimetic skills). In reproduction we expose ourselves, no one can hide, since what counts is not survival, but rather taking life beyond itself. Again, beyond the differences, be it a stickleback floating or a bird of paradise hopping, we all recognize and rejoice in reproductive exuberance because we feel that once again life has prevailed. Undoubtedly, in the reproductive process there are many aspects of sharing: the phatic performance, courtship itself, the sexual dimension, the unfolding of life seemingly out of nowhere, abiotic immobility with its miniaturized presence and its expectation of growth.

At first sight, animal-being means communicating, or being-for-communication: we feel united in the unavoidable and inescapable animal expressivity. Even in the deepest silence or the most absolute stillness, the animal encounter is communicative. We recognize ourselves in this eagerness for contact, no matter if this is achieved through the polyphony of the syringes or a vocal sequence, through mammalian horripilation or a mantis rubbing its wings, through a cuckoo exhibiting its yellow palate asking to be fed or a wolf licking its snout to provoke regurgitation, through a gardener bird putting together colourful stones or the dance of a bee, through the immense pheromonal vocabulary or urine-based directional signs. The animal cannot help communicating to its conspecifics, to heterospecifics, but also to itself, because in communicating it listens to and watches itself, relating to itself. To communicate means to declare that we exist. Communication cannot be without self-reference—a tacit satisfaction with our own expression. The innocence of communication, be it the roar of a lion or the hiss of a cobra, always moves and arouses a sincere and immediate harmony.

Humans and animals are equal in difference; the encounter with animal otherness communicates a sense of ontological solidarity that differentiates their shared experience of being animals from the encounter with a stone or a stream of water, and even from the majesty of a tree. Of course we admire the violence of a waterfall, a bright sunset, the magnificence of mountains or the vastness of the ocean, but we do not feel that sense of solidarity that comes from being under the same fate. A butcher feels a shiver when slaughtering a pig. This feeling can also be sublime: horror mixed with loosening and dislocation, since he knows that what is happening in that moment concerns him closely. Every slaughter is a self-slaughter. However, death is not the only thing that binds us together in animality, but also its opposite: that is, being-for-life, in which even death is no longer a limit or termination, but the fulfilment of a vital function, since only life makes death possible. In our common belonging to animality we find the pleasure of freely interpreting our *hic et nunc*, since animal-being means first of all owning the moment, having a here-and-now and not being absorbed by it, mastering what—from genome to parental care—was given to us, owning such endowments and not being in their power, in the short time we are given. Animal-being means having a fixed role in a play, but also being free to interpret it as we wish.

Therefore, if I recognize you in the animal that therefore I am (to put it with Derrida), our difference becomes colours, accessories, clothes we can exchange, words that can shorten the distance and allow for a common narrative through this identity exchange, in a sort of comedy of errors. Self-recognition means having a common basis on which we can lay the differences, in a trans-specific role-play that never takes our authority away, even in its dreamlike, stupefying, and disturbing nature. Variations on the theme of animality, biodiverse but with meta-predicative common grounds, represent the wood that can fuel human mimetic fire. To assemble them in one's body is not a difficult task, as they possess an intimate congruence: a perfect consistency with the ontological architecture of the flesh. To recognize oneself in the animal other means to decline the common grounds that tie us together, making me sympathetic with its presence, in a predicative elsewhere. It means to experience animal-being, and so to *be* in the deepest sense of the word, decentralizing myself from the specific predicates of human-being.

As we can see, this is far from a simple act of imitation: it is first of all a representative act. The flesh, understood in the dimensional sense of place or means of subjective expression, has a form, but underneath this form it shows common points (i.e. meta-predicates) that allow me to go beyond the form, liquefying it, so as to imagine a new shape. If epiphany is the act of imagining this new shape, zoomimesis is the act of taking on a new hybrid form: that is, the representation of the epiphany in our own body. Therefore, mimesis is not the duplication or the passive translation of nonhuman predicates into the liquefied flesh of man, or the transformation of the Epimethean predicate into a tool—copying nature through *techne*. Rather, it is an initiatory act requiring a long process of assimilation, but mostly of adaptation, in Piaget's words.[3] The encounter with the non-human animal is a slow and painful metamorphosis, one that excites us but also exposes us to vertigo, broadening our horizon but also increasing our vulnerability since it moves us away from our species-specific gravitational centre.

Mimesis is also an act of self-implementation, as the farther the orbit from its focus, the greater the probability of decentralization. But mostly, it is neither easy nor predictable, since it requires the ability to see the heterospecific being in an epiphanic way, not being entrapped in the phenomenon itself. In zoomimesis I must welcome heterospecificity and actively look for the keys to ontic integration, facing vertigo and finding new structures of mediation. First of all, I have to know how to go beyond the animal phenomenon—the simple manifestation of otherness—and bypass the risk of being swallowed by what appears to me as such, that is, the object facing my subjectivity, inevitably alien and thus usable. On the contrary, I have to feel it as a common living space, like a cosy home. This is animal epiphany, a manifestation that outclasses the phenomenon. Animal meta-predicates are its glue and its solvent: they have laid a thread of communion, allowing for the transfer of predicates and the ontopoietic splicing. Epiphany is inspiration and annunciation,

[3]J. Piaget, *Biology and Knowledge. An Essay On The Relations Between Organic Regulations And Cognitive Processes*, University of Chicago Press, 1974.

seeing the changes in ourselves and thus being able to look at our previous identity from a distance—the advent of the human that disavows man.

5.2 Inspiration and Revelation

There is no doubt that the non-human animals represent a paradox, as well as the recognition or the feeling of being bound in brotherhood. In other words, they represent:

(1) a difference or the alternative, despite the meta-predicative similarity, able to break the uniformity of animal-being in relation to the world;
(2) a breach in the projection of expectations, which is the magnificence of expression in difference—capable of arousing wonder and attracting attention with its non-correspondence to what is familiar to us.

As I have explained, the heterospecific being presents predicates of difference that are understandable because they are inserted into common metapredicative basis. In this way, the heterospecific is simultaneously the bearer of difference-plurality and of hermeneutical operators that can solve-explain this difference thanks to the concept of adherence and correlation to the environment. Therefore, to encounter heterospecific particularity means to have a guide that allows us to investigate the characteristics of existential problematicity of any environment-lifestyle, i.e. the challenges posed by that particular existential choice to the common coordinates of meta-predication, like eating, surviving, reproducing, communicating.

In this sense, animal otherness, through its predicates, shows the human being:

(1) that an existential alternative is possible even in respect of the meta-predicative needs of animal-being;
(2) how to achieve the mentioned existential alternative by providing predicates that are useful to these common needs, namely, to meta-predicates of consonance.

The paradox is carried out on different levels of evaluation and interpretation:

(a) predicative difference as such, as a multiplicity of meta-predicative variation, especially morphology, colour, expression understood as a plurality of life-styles, capable of arousing sillegic and exploratory interest;
(b) the correlation of predicates to the environment, in the two aspects a) of congruence to the context, i.e. performative specialization with respect to the peculiarities of failure, and b) of context mirroring, i.e. conformation both as complementarity to the context and as assimilation to the context (for example, mimetism);
(c) the understanding of the environment through the heterospecific predicate, or the use of the heterospecific sign (theriomorphism) to increase our knowledge of correlative mechanisms and thus enhance our predictive ability;

(d) the acquisition of new causal mechanisms or operational tactics that can make the targets pertinent to human beings easier to achieve.

The myth of Epimetheus can show us the wonder experienced by man in front of the predicative plurality of animal otherness. The Epimethean predicate is evaluated by humanism in terms of:

(1) compliance with a particular performative task, and in this sense it is considered as adequate to the performance and perfect in itself, that is, devoid of any performative counterindication unlike any human technique;
(2) rootedness within a rank, or having a specific predetermined position in a natural order that must be respected in its features, but also a counter-term of self-determination and freedom;
(3) overall balance of living beings, i.e. their mutual dependence in an all-encompassing structure that reveals the existence of a higher will capable of harmonizing the various instances, or the predicative plurality of different animals within a single well-balanced combination that can cushion its fluctuations. Thus, the figure of Epimetheus, even more than Prometheus, reveals anthropopoiesis to be the envy of animal virtues.

The encounter with heterospecificity takes place on two levels:

(a) the place of wonder, attributable to the animal paradox, which arouses admiration and interest; it is sacred and inviolable, comforting and reassuring, revealing a higher presence capable of establishing order out of chaos and of inaugurating the cosmos;
(b) the place of envy and emulation, attributable to the human desire to expand to new geographies, not only in terms of environment, but especially in terms of new demands coming from changes in society and lifestyles. The Epimethean collection can thus take the human being by the hand, thanks to metapredicative consonance, and lead it into the realm of the fantastic, of revelation, persuading it to transform dissonance into inspiration, into the inauguration of a possible dimension, into ontopoiesis. However, to achieve this goal man has to be able to "over-see", namely to see beyond the mere phenomenon in its foreignness and identify himself with that difference. Then, the heterospecific being fully takes on the face of otherness, since the human being recognizes the other as his neighbour, recognizing itself in the heterospecific metaphorical face. This is what I call "animal epiphany".

We often tend to bypass the epiphanic moment in the belief that the Promethean man either reaches new performative skills through solipsistic creativity or at least copies the Epimethean predicate, translating it into a tool through fire and *techne*. This goal is given as anthropocentrically inherent—for example, it is assumed that human beings have always dreamed of flying (their inherent *telos*), no matter if they saw flying creatures or not. On the contrary, it is believed that man, trying to reach his goal, goes in search of examples (something that flies). In this reading, the heterospecific contribution is either completely denied, as in the creative-solipsistic conception, or is confined to simple occasionality or necessity of exemplification. The aspect that is considered most important—the end of the technopoietic or cultural act, i.e. the autopoietic planning of the human being—is thus safeguarded. Thus, we do *not* interpret the emergence of human identity as a dialogical event that is not entirely confinable within human ontology. Only by turning Epimetheus into a foreign entity, thoughtless and akin to non-human animals, can the pure, exclusively Promethean human being be extracted. In this sense even the use of the Epimethean predicate as an example can be taken into account, as a disjunctive and paradoxically magnifying event of human autopoietic autarchy.

In humanism a sort of non-variation of the meta-predicative content is assumed: man is neotenous, shapeless, incapable, so he inevitably turns to *techne* for the expressive and performative functions that in other species are carried out by Epimethean predicates. The difference between man and animal is therefore identified not in terms of "predicative difference", but in terms of human lack of the predicates characterizing animals. It is in this absence of predicated that humanism grounds human freedom, the human autopoietic and protean potential, man's lack of a predefined rank, his possibility of being neutral and, therefore, act as measure of the world. So, man, while lacking predicative declinations, is paradoxically charged with metapredicative elements, and it is exactly in this surplus of exposure that the technopoietic tension arises, so that, for humanism, the goal remains firmly within the *anthropos*.

Relying on the tool, i.e. outsourcing the function, allows human beings to stay non-declinated, since the performative role is confined to the tool, i.e.to an external support, unable to affect human purity, which is fixed and overlaid like a glove. Hence the subsumptive view of the human being, able to swallow any external being without undergoing contamination. The Promethean reading of humanity is a swallowing vision that turns any external entity into material for the construction of its mythopoesis, disregarding any dialogic debt towards non-human otherness, in a tautological structure that always manages to make ends meet, since it is founded on the idea that everything that man has developed on his anthropopoietic path is the result of hard work and self-sufficiency, where non-human animals have only led the role of antagonists or counterparts. In this way, man presents himself as a heroic entity, brimming with goals to be realized, and humanity becomes a project of alchemical purification through the gradual emancipation from earthliness.

But—let it be clear!—in this reading the human "path" is not the construction but the extraction of humanity, i.e. the realization of what was already inherent to it. The here proposed epiphanic interpretation of heterospecificity—which clarifies the arguments I have already exposed elsewhere[4]—calls into question this very instrumental reading of non-human animals and its humanistic corollary—especially the concepts of inherence of purpose and autopoiesis in human realization, and, mostly, the idea of extracting the human being through processes of exemption and purification from external contamination.

[4]R. Marchesini, *Post-human. Verso nuovi modelli di esistenza.*

The epiphanic appearance of the non-human animal does not have to be considered as an example of useful or viable modalities—that is, heuristics that can be Prometheically translated into *techne*—but rather as the moment of man's ontological shifting due to an external referent that is no longer a tool, but a guide. Epiphany is realized at the very moment in which the animal other loses its connotations of strangeness and becomes the companion to another dimension, or a dialogic entity, being finally recognized in its subjectivity. Epiphany is precisely the annihilation of the heterospecific's manifestation as a purely phenomenal content: man can over-see animal otherness only when he stops being dazzled by the animal as such. Only then can the heterospecific being truly be-with, and thus paradoxically be able to show its predicative difference.

Suddenly, removing the mask that estranged it and fully exposing itself metapredicatively, beyond the appearances of specific predication, animal otherness is able to speak for itself. Thus, it can reveal its predicative connotations not as purely phenomenal elements or elements of distance and disconnection, but as possible existential dimensions, open to human beings. In this sense animal epiphany is a revelation, like seeing for the first time a bird flying or a fish swimming, seeing it in a dimension that not only exceeds von Uexküll's functional connection (the flower as a means to acquire honey) but also Heidegger's objective evaluation (the flower itself), becoming a metamorphic a space of transcendence. The flight of a bird is my opportunity to transcend the limits of my phylogenetic heritage and strive for another dimension, which is over-human as it comes from a process of hybridization with otherness that can change not only my tools of intervention in the world but my whole perspective on it, modifying the very meaning of my existence in the world starting from my own *telos*.

Animal epiphany means going beyond our own nature, not to contradict it, not to deny it, but simply to give it a different scope and a metamorphosed internal organization or form—which is frightening at first and gives us a feeling of bewilderment and vertigo. In short, it evokes an aesthetic feeling that is profoundly different from narcissistic beauty. What does man feel when he suddenly realizes that flying is possible? When he imagines himself flying? When he feels bewilderment coming from this revelation? He feels enthusiasm, namely a tension vibrating throughout the body, and ecstasy, or the body's loss of gravity, the feeling of flesh liquefying and an unrecognizable face, a therianthropic profile projected by epiphany. Man sees a new dimension: the image of an over-man that has now accepted his human identity. The human being has never seen itself simply as a man, because its first gaze in the mirror was reflected a heterospecific nature. A modified image was reflected, as the animal mirror is not neutral and it charges the human face with non-human predicates. I believe this is the reason why we cannot speak of the human being without referring to animal otherness, since it would be like trying to see ourselves without using a mirror.

Animal epiphany means going beyond our nature and projecting ourselves into the animal body, overlapping with its predicative architectures and flying into its flesh, welcoming the heterospecific heartbeat inside our body, in a real rite of possession, in a trance that dispossesses the human being of the full ownership of its expressions.

Animal epiphany shows a possible world, but mostly it modifies our perceived destiny, the horizon of the goals that the human being can reach. From epiphany we emerge irreversibly changed: it is an initiation. For this reason, we cannot easily free ourselves from the therioantrophic body we inhabit—which is in us, who are in it—because from that moment the heterospecific being is no longer other-than-myself, but rather other-with-myself.

Once we feel the epiphany of flight, we become irremediably bird-like, and the our preservation expressive dimension can no longer be the one proper to man, as an entity phylogenetically defined and ontogenetically realized into the previous parameters. An ornitomorphic being has different needs, perspectives, and purposes and has by now metastatically spread to all the clefts of the human flesh; it has reorganized them according to morphopoietic schemes that have crossed the phylogenetic boundary. An agreement between the human being and animal otherness has been established, one that leads humans towards a zoomimetic expression, namely towards an ever more perfect adhesion to the hybrid dimension introduced by animal epiphany.

5.3 The Theriomorphic Sublime

Mimesis, according to a well established tradition that goes from ancient Greece to Romanticism, is often presented as a simple, if not passive, imitation of nature; a copy of a world that, according to Plato, is itself reflected and therefore doubly spurious and misleading. Yet, if we consider authors as far apart as Aristotle, Lucretius, Bruno, and Goethe, we'll see that the observation of nature has always corresponded to the contemplation of form, a form that presents specific internal or holistic constraints, as Gestalt theorists or new Evo-Devo explorers have noted.[5] Form gives us a living image that meets specific expectations such as symmetry, proportionality, disposition according to the golden ratio, certain chromatic combinations, prototypical structures that, as pointed out by Wilson in ‚*Biophilia*[6] exhibit aesthetic characterizations that are innate to the human being. The species-specific predicate confirms this expectations in its correlative consistency, as it can reflect the environment and lifestyle in the animal body: the prehensile hand of a primate mirrors the shape of the branch just as a fin reflects the viscosity of water and a wing reflects the resistance of gravity and updrafts in the air.

---

[5]See S.B. Carroll, *The Making Of The Fittest: Dna And The Ultimate Forensic Record Of Evolution*, Norton, New York 2006 and A. Minelli, *Forme del divenire. Evo-Devo: La biologia evoluzionistica dello sviluppo*, RCS, Milano 2009.

[6]E. Wilson, *Biophilia*.

The beautiful is the feeling of finding ourselves in morphological cohesion to otherness, recognizing that the meta-predicative root of consonance is perfectly conveyed in its specific performative variation, and therefore it can reflect a body-image on the world because it is itself the reflection of a specific dialogue with the world. The contemplative observation of nature soothes, languishes and cheers us up; it always has to do with beauty, even when it shows us vastness and magnificence. The terror-tremor of the sublime, on the contrary, is always the result of a different evaluation, that is, of feeling decentralized, shapeless or unstable in form. In the image crystallized in its form through declination, be it majestic, curious, unexpected or paradoxical, the human being feels wonder, as emotional and cognitive opening, as the preliminary question to knowledge. The sublime is its very opposite: a disorienting sensation, the impossibility of remaining anchored to a form, the dizziness of instability and heteronomy, the stupefying ecstasy of explosion and annihilation. The sublime is the moment when the contemplation of the heterospecific as a phenomenon changes into a trans-specific epiphany. The sublime is therefore every threshold crossing, every desecration, every risk of *apeiron* or the bewilderment in finding yourself dismembered and diluted into the non-human animal.

And yet, as a long sequence of founding myths shows us—from Prometheus, inspirer of *techne*, to Adam, flattered by knowledge, from Pandora's inability to resist temptation to Daedalus' ambivalence—man is essentially *hubris*, so that the sublime, as Baumgartner would put it, turns into creative knowledge that shifts the focus on new levels of reality through hybridization. The epiphany opens up an over-human space that envisages the theriomorphization of human flesh: it is there that mimesis operates. By wearing the theriomorphic mask, the human being embarks on its anthropopoietic journey. Therefore, mimesis requires the courage to face sublime vertigo, to accept decentralization as an enhancing opportunity for representation and as a space for freedom, in order to strengthen the awareness of our own identity. Just as in the pedagogical "acting for self-understanding", in the mimetic act the human being does not merely copy, but comes to terms with itself, confronting its integration issues and the other's efforts of interpretation. Mimesis means going beyond oneself.

When the human being is called to reproduce the species-specific predicate, it must incorporate it through two operations of form liquefaction:

(a) the sublimation of the meta-predicate from the anthropocentric declination: a process that is inevitably partial and presents differing degrees depending on the levels of anthropomorphous-critical recognisability;
(b) the overlapping of one's partially sublimated body and the heterospecific with a hybrid and deformed outcome, opening up new existential dimensions, which in turn require an active representation on the part of the subject. It is in this double liquefaction of the accredited form that, from the experience of the beautiful, we come to the sublime. The human being is inevitably faced with the unknown new possibilities opened by epiphany, but from a weakened and decentralized position. The sublime is therefore the experience coming from the

transcendence of the form, from the sudden confrontation with the *apeiron* of meta-predicative sublimation and the contaminating flow of other predicates; it is a feeling of *hybris*, desecration, breaking a taboo, needing reconciliation. It is the sudden empire of dissonance.

Mimesis is an active re-proposition: it is represented in a form that is free albeit contained within a stage of flesh. It is the emergence of an ontological novelty that can be reduced neither to the artist's autarchic creativity nor to the mere duplication-copy of animal otherness. The form is transfigured, the image is moved and for a moment there is neither a man nor a bird, but only a man-bird, whose existential dimension transcends both. Man can no longer identify himself as fully human: that space is not enough for him—it is narrow, lower, ancestral, that is, anthropocentric—but at the same time being decentralized feels like having shipwrecked. Anthropocentric space cannot represent the bird-shaped man since its ontological perimeter cannot contain it any more, hence the experience of the sublime as a feeling of unsteadiness coming from dislocation. When the coordinates of spatial allocation disappear, delirious irrationality takes possession of the human spirit, arousing the ambivalent feeling of the possessed self. The possessed is controlled by the otherness inside itself and, at the same time, it is projected into the innumerable reflections of otherness, like Dionysus. This is the loss of the unity the rational, ordered and ordering self. The mimetic act breaks all boundaries, putting the human being in front of the infinite—and not vice versa, because the human never gravitates within man but is a principle of decentralization. The human being is therefore the way in which man flees from himself: anthropogenesis means becoming *hospes* in the presence of otherness.

Being decentralised is the basis of the sublime: it is the feeling of being too far away from home, of flying into a bird-shaped body, feeling we are not in control of our mad ascending, as the endowments that should guide the kinetic declination are anthropomorphic predicates of no use for flying. Therefore, Il being decentred is not a starting condition proper to the human nature; rather, it is the outcome of a mimetic process beginning with epiphany, or when the human phylogenetic body melts to give life to a symbolic hybrid body that transcends the former because it has been infected with the predicates of difference. When the human being becomes aware of its inspiration—or rather, precisely through this slow process of perception of the theriomorphic emergence—the representative process begins: the human being starts to understand what it means to see itself in the therianthropic, hence the zoomimetic act begins as a factual re-proposition of the epiphany. However, hosting otherness or being hosted by it, in the twofold hermeneutics of the Latin word *hospes*, means precisely being decentred and fleeing from one's original condition towards a transcendent condition. In order to find the "human", we need to move away from man, from its phylogenetic perimeter and, mostly, we need to stay in an ever-negotiable context, which never reaches a definitive settlement. Mimesis is tension-sufferance making us part of something bigger, something that drowns the self, making it shipwreck it its infinite vagueness.

Hybridizing factors stem from a Promethean component of mimetic tension and also from the meta-predicative recognisability of Epimethean contents. There is a strong relationship between animal epiphany and technopoiesis, since, through *techne*, Epimetheus gives life to new existential dimensions in human ontology. *Techne* is the new horizon of therianthropic humanity, the translation of the non-human animal predicate into human ontology, the transformation of human flesh along a new declinative chassis and the reorganization of philogenetic content by a contagious otherness. The telescope does not project the eye into the infinite; rather, it makes it shipwreck, since technology is not at all ergonomic and does not follow man's predicates, but it calls them into question like un virus that, breaking into the cell, reorganizes it to its own benefit. The sublime is the perception of the instability coming from being decentred. If, on the on hand, the human horizon broadens, on the other, man's home gets smaller. So, the sublime is not infinity as such, but the abyss that appears in the human soul whenever man discovers the multiplicity of possibility and whenever the certainty of his judgement and his identity ontogenesis is questioned.

If Prometheus is the driving force of mimesis and Epimetheus is the site of the epiphany (the two dialectical reasons that lead human ontology to liquefaction), we must admit that the human being is neither the emanation of man nor the gravitation into man, but the opposite: namely, the event that decentralizes and allows for both the ontopoiesis—as the free construction of the identity ontogenesis—and the ability to see oneself, in the identity tripartition that occurs between man, bird and man-bird. Faced with monstrosity, needless to say, we feel affinity; we are certain of the infinite teratomorphic endeavour with which we have built our own identity, whenever we had to look like heterospecifics—wearing animal skins, shaking our body like in gruiformes and Paradise birds, scarifying and tattooing our skin to give it non-human consistency and colour—in order to recognize ourselves as human. *Hybris*, *Apeiron*, delight? Mimesis is the outcome of multiple motivational chiasms, never a primary emotion; it is more of a composite feeling coming from interest, attention, explorativity, enthusiasm, syllegic tendencies, envy, emulation. And if this has any ethological and psychological plausibility, then we are destined to the sublime or to continuously overcome ourselves, suspending our self in the vertigo of possession and centrifugation.

Mimesis lies at the heart of this process, capable of turning any phenomenon into an ontological epiphany: a flying bird is no longer just a reality "as such" in a Heideggerian way, but a space of possibility opened up through crossing the threshold and accepting the risk of *apeiron*. In a bird's flight man sees himself flying, an overlapping between the two bodies takes place, and the human being emerges when man (as a phylogenetic entity) becomes animal otherness, is welcomed by it and welcomes it into his body. To see a flying bird does not mean to learn how to fly, but to acquire a new existential space, to understand that "flying is possible", to build a new symbolic body that transcends us and, broadening our domain, makes us dizzy. The sublime means to experience flying even before building a device capable of turning the Epimethean predicate into *techne*: it means to imagine-feel the dizziness of flight as space of possibility. Mimesis turns every otherness into an epiphanic counterpart, able to enter the anthropomorphic body and modify it to create a hybrid, heteroreferred and heteronomous dimension.

This leads us to think that the human being is the continuous sublimation of man. The moment when decentralization inaugurates a new level of reality, it also calls the conformational Apollonian into question, so that the reflection on the sublime emerges whenever a strong decentrative process begins. So, we have to better focus on the narcissistic wound that underlies the experience of the sublime mentioned by Saint Girons in his essay *Fiat lux* when referring to the evolution of knowledge in the Seventeenth century.[7] It is not so much the human being that is hurt, but man in its anthropocentric heritage and ideology. The hybrid dimension takes us to the sky, but we are not equipped for it as an eagle or a swallow, so we cannot superimpose our body to the phylogenetic symbolic body that has emerged from mimesis. In mimesis a new dissonance opens up between the teriomorphised symbolic body and the body as a phylogenetic entity. This is the dimension of the sublime, which amounts to discovering the vulnerability of our phylogenetic body compared to the hyperspaces of the symbolic body. The sublime is the experience of weakness that we feel not in the presence of nature's greatness, but in the relationship between the phylogenetic and the symbolic body—a gap that widens when hybridizing events increase in number. Then, man must rely on the hybridizing otherness, not only to fully realize his new existential dimension, but also to gain new spaces in the sublime.

Therefore, mimesis is the ability of a human being to get in tune with the outside world and make it its own through a representation focused on the body. In fact, the body is the place of hybridization where epiphany is revived, not in the mere imitative aspect, but in the most dynamic, dialogical, open space of representation: the body is therefore the stage and the laboratory where the epiphanic event is represented and the experience of the sublime occurs. Mimesis is thus the foundation of the human being, allowing us to deny a *juxta propria principia* foundation through an autarchic-emanative explication: by means of zoomimesis we acknowledge the dependence of anthropopoiesis on non-human referential intervention. The work of art comes to light as a primitive form of mimesis: not a mere replication of reality in an anthropomorphic game of mirrors, but rather a trans-valued representation of the body. Every work of art is mainly painted on the human body—the canvas is a surrogate for it. It is in projection and decentralization that mimesis inaugurates the experience of the sublime, transferring the existential dimension of the human from the phylogenetic to the symbolic body. On the other hand, it is precisely in this break from the phylogenetic anchorage that tremor occurs, when the human being realizes it needs otherness in order to achieve its symbolic body.

[7] B. Saint Girons, *Fiat lux Une philosophie du sublime*, Quai Voltaire—La République des Lettres 1993.

5.4 The Relationship Between Zoomimesis and *Techne*

As I have explained, mimesis operates a slippage, it opens up a space of escape, turning the construction of individual identity from the autarchic act of a single species (ontogenesis) to a transversal process open to the contributions of other species (ontopoiesis), starting a non-anthropomorphic modality of projection into the world: *techne*. *Techne*, then, is not the result of human creative solipsism, but the dialogic tension that leads human beings to transcend themselves through the ontological hybridization with other species. Therefore, technology is nothing more than the mimetic representation (in a representational, not imitative sense) of the Epimethean predicate (of heterospecific origin), which is introjected-hosted in the new theriomorphic condition of the human being as a result of the epiphanic revelation. If the epiphany is the moment when the human being discovers the theriomorphic possibility in the encounter with the non-human animal, imagining itself as theriomorphic, and if mimesis is the act of representing this new being, which is neither man nor the heterospecific's copy, *techne* is the moment in which this over-human being is confronted with its new ontological status, namely, with its irreversibly changed perspective.

It is now clear that, in a zoomimetic interpretation, *techne* is not the creative and autarchic predisposition of the Promethean man; rather, it is the outcome of a Promethean and Epimethean dialectic. In other words, *techne* is the result of:

(a) the tensions of the Promethean human being, who is projected into the world and is fascinated by multiple biodiverse predicates, underneath which it still manages to recognize itself;
(b) the inspiration and decentralization coming from animal epiphany, when the overlapping with otherness brings about new opportunities and risks to its existential horizon.

Techne is therefore the result of this dissonance creating fibrillation in the creative process of somatization and psychedelia, an enlargement of body-consciousness that becomes a metamorphosis of the human interface. Thus *techne* is a new way of communicating with the world: a different intentional structure that brings me to the world through different coordinates.

The Epimethean contribution can be summarized as follows:

(1) the ability of heterospecific otherness to dissolve man's ontogenetic autarchy through the multifarious' appeal on our species' motivational structures;
(2) creating a connection through the revealing property of our common animal being; making itself recognizable in meta-predicates;
(3) envisaging in an epiphanic way other possible declinations of human-being and representing them through mimesis (meant as a hospitable active action, that is, animal-becoming, reproducing heterospecific otherness in our bodies);
(4) giving life to new existential dimensions, creating a second interface or intentionality that breaks the anthropomorphic projective scheme.

Returning to the example of the man-bird epiphany, *techne* is how the bird-shaped being experiences the world through its new existential dimension. *Techne* is the renewed *Umwelt* of the human being, who is not only phylogenetically man, but human and therefore able to bring out new and unusual environmental markings and operational styles: namely, techniques. If this is true, we need to change the very way we interpret *techne*: no longer an instrumental function, ancillary to human purposes, Promethean essence and disjunctive banner of the self-sufficient human being, it becomes a way to strengthen the dialogue with Epimetheus, a connection to the non-human and an exchange of predicates—that is, exactly the opposite of autarchy. Therefore, *techne* means building new interfaces between man and external reality; it is a communicative modality in a context-world sense, where discerning perception from operations makes no sense, since it refers to the existential totality of Being, which is decentralized from its phylogenetic body.

Techne is an intentional, i.e. referential, function, just like the phylogenetic *Umwelten*, which we contradictorily take to be closed structures, soap bubbles, when in fact they are evolving structures, since their very referential task implies evolution. Both in plant structures and in a neurobiological plexus, to express also means to grow; so, if perceiving is already operating, to express oneself (which covers both things) already implies changing oneself. Thus, every *Umwelt* is recursively a condition destined to self-transcendence. The referential function not only means that we enter the world, but also that we transcend ourselves through our experience of the world. In this non-traditional sense, if we take into account its intentional capability, there is no difference between the concept of *Umwelt* and the concept of *techne*. The difference lies somewhere else: while an individuality that is hardly released from species coordinates always emerges from the body's phylogenetic evolution, the advent of *techne* creates a decentralising acceleration that cuts the umbilical cord, opening up an ontological non-equilibrium that makes the human being increasingly open to heterospecific contamination.

Techne as a form of communication with the world is both the result and the driving force of decentralization. We can define zoomimesis as an accelerator and amplifier of the ontogenetic process, i.e. of the emergence of individuality from its phylogenetic legacy. Decentralisation is therefore contemplated, albeit with restraint, in any ontogenetic process of realization of an individual identity. In the evolutionary emergence, the individual always transcends the phylogenetic dimension and yet this decentralization, capable of realizing subjectivity, always remains within the coordinates of the species. The mimetic act, which takes place after the epiphany, opens up a new situation. In fact, when animal otherness is revealed not as other-than-oneself but as "evolutionary space" (another way of calling animal epiphany) whenever the human being feels itself in the reproposition of the heterospecific predicate within its own body—what I have called zoomimesis—then a new context-world emerges for the subject, i.e. a new *Umwelt*, that is, a new individual expressivity, both as interpretation and as operativity in the world (what I have called *techne*). If ontogenesis itself neither question anthropomorphism nor creates a rupture between the phylogenetic and the symbolic body, the mimetic act, on the contrary, opens up a gap, transforming the human being into an entity that projects itself beyond the anthropomorphic. But this new dimension lacks balance, resulting in an increased adherence to otherness and a greater dependence on the non-human.

The sublime is this feeling of losing control, the bewilderment upon realizing that we are no longer holding the reins of our journey, the feeling of having shipwrecked. So, it is correct to talk about techno-science, as with the advent of *techne* both the epistemic dimension and the human *telos* have changed. This is also the reason why science has always been perceived as a slippery slope. In every age, from Galileo to Darwin, scientists have gone against the tide, exposing themselves to the risk of being mocked and discredited. The reason for this is very simple. Science is counter-intuitive: a rupture with the anthropocentric epistemology, insulting the existing paradigms and forcing its proponents to self-exposure. Science is always an act of *hubris*, the liquefaction of every *status quo*, the resetting of the system, blasphemy towards sacred values. Therefore, we should not be surprised if the most invasive operation of mimesis, i.e. the singularity of the seventeenth-century scientific globalization, has not only inspired the human being, but has also assumed the ambivalent profile of strength and vulnerability. Science as techno-mediated mimesis decentralizes the human being from the reassuring perimeter of anthropocentric epistemology, showing the counterintuitive resistance of the real, as shown by the Copernican paradigm and Lavoisier's theory of oxygen, downgrading the intuitive to the rank of naive physics, i.e. a lower interpretive domain. Science challenges anthropocentric ideology, shattering the Vitruvian Man, since man can no longer be the measure and subsumption of the world.

Man discovers the infinity of the formless within himself and in this sense, only in this sense, we can say that the Copernican infinite produced two centuries of disease in Western culture, which only Kant attempted to reconcile. Charles Darwin can be considered another episode in this counterintuitive saga: the last anthropodecentralizing addition disrupting human certainties and making the anthropocentric universe fall apart after it had already been undermined by a surplus of infinity. But, then, what was the outcome of Darwin's perspective? Not a gap, as was first envisioned due to the idea of profanation implied by the concept of common ancestry; rather, the radiation of a thousand imperceptible stains that led in the twentieth century to a far more blatant erosive action than the innocent reference to our baboon grandfather. The most important dichotomies with which man had built the humanistic system—such as the couples nature-culture, natural and artificial, man-animal—fell like bowling pins after being hit by the well-aimed "evolutionist blob". We soon realized that we had to deal with a far more insidious and degenerative virus than the first blatant access, which was easily amendable with containment operations, as shown by the (scarcely successful) proposals by Helmuth Plessner and Arnold Gehlen.

Of course, if the Darwinian problem were merely a matter of relationships and family trees, with a little trick like that attempted by Patrick Tort, everything could have been preserved. In fact, it turned out that the baboon grandfather cannot bewilder us: it can only stir our haughty conceit and, at most, our confusion. Yes, perhaps it had a few more stains, but the humanistic robe—reassuring in its six centuries of experience—did not necessarily have to be thrown away. It was even still possible to raise the stakes, as Heidegger did in his *Letter on Humanism*.[8] But the English heretic is not responsible for this, and maybe the core of the Darwinian proposal does not constitute the only reason of the paradigmatic metamorphosis that began just after the publication of *The Origin of Species*. Darwin planted his seeds in the fertile ground of the Twentieth century—the age of machine domestication and the panopticon over the *bios*, the time of information technology and genetics, of animality and ecstasy, of the sublime of biodiversity.

This is a poetic reversal: from a human body dictating the metrics of the world, we move to a non-human constellation that scans human morphogenesis. Hence Matthew Barney's theromorphies in the alchemies of *Cremaster*; the non-human heroes imagined by Philip Dick; metamorphic superheroes—from Batman to Spiderman—who are ambiguous carriers of strengths and weaknesses; the animal cyborgs that are the new extenders of a human manifesto that is also the inventory of the new *cahiers de doléance* (as Donna Haraway calls them), and Miyazaki's fluctuating and unstable metamorphosis. What is affirmed is no longer the stigma of the ancestral or the regressive—the beastly Mr. Hyde, Truffaut's wild boy, Conrad's impenetrable Kurz—but the posthuman sublime, reverie of the Nietzschean dream purified of all forms of desire for power. It is the primacy of the hybrid and the mutant, so that the elevation takes on a trajectory that is the more oblique the more it is unstable. The Vitruvian man is therefore dissolved in a primordial soup that multiplies identities and makes them transient, underlining the importance of dialogue. It is not the end of the human being but its assertion on man.

In the posthumanistic sublime, the ontopoietic event is always heteroreferred—*dialogo ergo sum*—which does not imply a passive assumption of the form imprinted by the world, but rather a morphopoietic emergence, attributable to the ability to recognize one's own body as the scene of representation in which the human being and otherness, placed in a hybrid interface, define new predicates. The post-humanistic over-man is therefore not the autopoietic and disjunctive entity of Pico's manifesto, but an entity that builds its own symbolic body through conjugation with otherness, which reaffirms its reliance on dialogue with non-human animals. So, it is undeniable and understandable that these unusual figures generate new experiences of the sublime; and yet, since infinity existed before the scientific revolution, similarly the hybridization with the non-human being is not produced by Darwinism, but revealed by it. We have always been hybrid, perhaps even before our ancestors re-negotiated our morphopoietic threshold by chipping a piece of flint.

[8] M. Heidegger, *Letter on Humanism*.

Chapter 6
Steps Towards a Philosophical Ethology

Over the past century, the research on animal subjectivity has had to face several obstacles. On various occasions, there have also been waves of anthropomorphism and projective (more than continuist) interpretations of the animal behaviour followed by reductionist or mechano-morphic reactions. Some authors have applied the reductionist explanation to both the human being and other species—think of Skinner's behaviourism or Wilson's sociobiology[1]—filling humanists with indignation and arousing fierce criticism in most of the philosophical and anthropological community. Others, instead, have preferred to follow Descartes' dichotomy, which allows one to explain animal expression through mechanistic coordinates without harming the sovereignty of anthropocentrism.

To consider animality as *res extensa* means to transform its expression into a sum of automatisms that are activated either by external stimuli or by inputs directly derived from somatic physiology. The explanatory mechanism offers significant

[1]At a first sight, the two traditional accounts—as expressed by Skinner, on the one hand, and Tinbergen, on the other hand—describe mutually incompatible explanatory models: (1) according to behiviorism, the animal works as a trigger machine, (2) according to classical ethology, the animal is a pressure cooker. As we can see, these two explanatory models are predicated on a basic assumption that is not called into question, namely that the animal is a machine. The scientific explanation of animal behavior, in searching to define "how the animal machine functions" and restricting itself to defining features of animal mechanisms, has never problematized the Cartesian paradigm of the animal as automata. The only thing that has been problematized is whether that mechanism is an instinct or a conditioning. As in Kuhn's view, science has only chosen the type of machine, remaining well anchored to the philosophical paradigm. For this reason, I believe that only a philosophical approach which problematizes the basic paradigm—namely the Cartesian idea of animal automata—can really overcome ontological anthropocentrism: unless the human being is viewed as nothing but a puppet deterministically moved by strings. The main principle is refusing mechanism, or, to put it differently, acknowledging animal subjectivity. In order to do so, a paradigmatic alternative to Descartes must be found. Cf. B.F. Skinner, *Science and human behavior*, New York, Macmillan, 1953 and E.O. Wilson, *Sociobiology: the New Synthesis*, Cambridge, Mass.: Belknap Press of Harvard University Press, 1975.

R. Marchesini, *Over the Human*, Numanities - Arts and Humanities
in Progress 4, DOI 10.1007/978-3-319-62581-2_6

methodological guarantees, because: (a) it answers, or rather seems to be devout, to Ockham's razor, redefined within Morgan's canon of parsimony; (b) it does not call into question external or tautological entities, such as the *res-cogitans* or the *homunculus*, which avoid the test of scientific explanation; (c) it agrees with the epistemological principles of description, falsifiability of the explanatory hypothesis, and causal modelling; (d) it is perfectly consistent with the canons of Western culture that radically separate animality from the human condition.

During the twentieth century were thus developed some very convincing interpretative coordinates that seem to have put a tombstone over the theme animality:

(1) the animal expression found a final explanation in the set of automatisms; some driven by phylogenetics and derived from the innate (instincts), others emerged from ontogenesis following the law of effect, and thus learned by the individual (conditioning);
(2) human beings distinguish themselves by separating or emancipating themselves from the animal condition; the human evolution is not analysed in terms of adaptive specialization—that is, in a specific sense—but rather in terms of lack of adaptive declination—that is, the human being is special, free from automatisms and thus autopoietic. This interpretation separates subjectivity from animality so as to connect it closely to reason, language and consciousness.

This "speciality"—rather than "specificity"—of the human being has given rise to different interpretations, from Martin Heidegger's ontological difference[2] to Arnold Gehlen's adaptive deficiency in line with the humanistic paradigm.[3] However, these interpretations are all essentially based on Descartes' dichotomy, which aims to lead animality back to the determinism of the *res extensa*, from which humans distance themselves for their meta-predicative abilities. According to this interpretation, humans were not born from a particular declination of their animality but by emancipating and distancing themselves from that dimension. Darwin's theories have been adapted to this interpretation as well, so that the human condition is not the outcome of specific adaptation or predicative declination to a certain performativity: it is rather due to a sort of reverse effect.

Animality thus falls into the realm of determinism, that is, the mechanical translation of behavioural automatisms that are necessary/exhaustive to explain animal expressiveness: the animal is the sum of its automatisms and its expression

[2]Martin Heidegger understood this aspect very well when he realized that the predicative explanation, still in vogue among the old humanists fascinated by the plasticity of the Vitruvian model, was not the crux of the matter because the difference between the human being and the animal otherness was ontological and thus meta-predicative. The animal machine can perform actions, take part in some functions, be comprised in a period of time, and so on. But if/until it remains a machine, it will never be really present, because a machine is in an isochronal state and does not own a hic-et-nunc. Cf. M. Heidegger, *The fundamental concepts of metaphysics: world, finitude, solitude*, Bloomington: Indiana University Press, 1995.

[3]A. Gehlen, *Der Mensch; seine Natur und seine Stellung in der Welt, Bonn*, Athenäum-Verlag, 1950.

is their functional outcome. Cognitive ethology—a branch that attracted a multitude of authors, from Kohler to Griffin[4]—tried in vain to introduce fragments of subjectivity into the animal-machine by recognizing in it levels of intentionality that, albeit lower than those of a human being, can bring out a glimmer of subjectivity by reflectively and explicitly taking care of some processes or psychological states. Subjectivity is thus related to the knowledge of states such as emotions, intentions, the creation of goals and strategies (that is, behavioural molarities), the ability to find solutions through insight, the creation and use of tools and so on. In this context, one is subjective insofar as one is aware of what one feels or expresses through behaviour.

This theory shows very well some of the main events of animal psychism: evaluation, judgment, decision, planning, simulation and creativity. However, I believe that it does not help understand the foundations of animal subjectivity, because, as long as one models the endowments as automatisms—that is, as expressive cogencies—there is no room for behavioural flexibility. A machine endowed with consciousness can only take care of the mandatory nature of its mechanisms. Therefore, to add a consciousness to the traditional model of the *res extensa*, which would only restore Descartes' dualism by trading the metaphysic *res cogitans* with an informatics or cognitive *res*, is not a solution. What really matters is to question the very principle of *res extensa*, that is, the modelling which animality is based on.

The ethological research, as well as the cognitive one, has analysed predicates in detail, without questioning Descartes' paradigm of the "animal machine". Instead, it has simply proposed, each time, a particular model of machine: (a) a psycho-hydraulic machine in classical ethology; (b) a cybernetic or feedback machine in behaviourism; (c) an informatics machine in cognitive proposals. My goal is therefore to question the basic paradigm, that is, the animal ontology, in a conception of existential—and therefore ethologic—plurality. In order to do this, I will start from a meta-predicative and therefore philosophical reinterpretation of the animal.

First of all, some important points have to be developed:

(1) the basic model that explains the individual's natural and acquired endowments must be redefined;
(2) the concept of subjectivity must be reintroduced, without falling into the tautological trap of consciousness or other dead ends that would lead to an infinite regress;
(3) one must explain the emergence of the psyche rather than the total state of the body and its relations with the outside world;
(4) the concept of existential plurality must be highlighted, avoiding the anthropocentric tendency to consider the human being as term of comparison so as to grant the heterospecific a certain form of presence.

[4]W. Köhler, *Gestalt psychology, an introduction to new concepts in modern psychology*, New York, Liveright Pub. Corp. 1947, and D.R. Griffin, *Animal minds: beyond cognition to consciousness,* Chicago: University of Chicago Press, 2001.

It should be said from the outset that one can speak of subjectivity only by acknowledging an "ownership" of the behavioural endowments: the individual uses the endowments; they do not control him. Otherwise, if the endowments were able to directly explain animal expression, there would be no room for the subject to appear. Moreover, one must wonder whether a different paradigmatic structure could better explain the animal behaviour:

(a) by resolving some inconsistencies present in the current explanatory models;
(b) by avoiding to call into question entities that are metaphysical or that cannot be explained clearly;
(c) by respecting the principle of parsimony.

6.1 Shifting from an "Automatism-Based Model" to an "Instrument-Based Model" in Order to Explain the Endowments

To be able to talk about true ownership, one must first of all modify the model that describes and explains behavioural endowments. The mechanistic model is based on the concept of "automatism", whose structure is as such able to explain the expressed function. The automatism transforms the behaviour in a sort of imperative and substantially deterministic reflection in the function produced, where an access lock for the stimuli (both internal and external) corresponds to a behavioural mechanism, such that each input matches a precise output. Both the key signal describing the instinct and its elicitative processes, and the structure of classical conditioning taking part in the dialogue between stimulus and response abide by the same basic requirement: the mechanism exhaustively and cogently explains the behaviour. The automatism thus becomes the explanatory paradigm of both the phylogenetic heritage and the learning processes.

Complex behaviours can also be explained through this modelling, considering the single automatism as nothing but the atomic unit from whose composition in sequence—the behaviourist chaining—or in other correlative, synergistic or antagonistic forms, any complex behaviour may result. To consider behaviour as the sum of automatisms activated one after the other leads to an analytical explanation of animal expression, where each molarity is led back to molecularities combinable with one another, in a context of expressive, rather than virtual, units. In order to understand this explanation, it may be useful to picture dominoes in which each piece can be a meaningful answer to the incoming wave and a stimulus for the next one.

A modelling that does not take place through virtual connections but through a succession of expressive units does not take into consideration the organization of the neurobiological system that has a connective-systemics, rather than associative, structure. As soon as one considers the mental state as a particular systemic activation of the network—that is, as one of the possible "states of the network" given by a certain connective activation of its structures—the model shifts considerably. Without considering every single behaviour as a specific expressive domino structure, the functional unit could be seen as a "group member". However, its outcome derives from the systemics. Supposing that each functional unit allows for a different expressive conversion that depends on the systemic configuration activated at a particular time, one is faced with a model that is more parsimonious and more responsive to neural connectionism than to those autonomous units called associations.

The functional systemics thus shows endowments that are more reminiscent of a map schema than a domino game—that is, their structure can take different functional configurations, implying a series of possible expressive paths—and thus gives rise to a multiplicity of functional outcomes. So, thinking about the difference between a domino-based model and a map-based model, one can immediately realize that the functional meaning shifts considerably. While the domino-based model exemplifies the functional automatism (perfect translation of unavoidability and determinism), the map-based model shows endowments as tools. This shift in the "explanatory model" is quite significant. In the automatic model, the ratio of structure (what it is) to function (what it does), can be represented as 1:1—that is, "what it is = what it does", so that the function can be obtained from the given structure. In the instrument model, the ratio of structure to function is 1: infinity—that is, the instrument-endowment (a) arranges itself, (b) gives competence, (c) becomes available to the function but does not determine it.

To consider the systemics responsible for the specific functional translation of the endowments means to recognize a surplus in the explanation of the expressive component that cannot be directly traced back to the endowments. This does not mean to deny the endowments a performative characterization, but to think that the endowments develop a range of possible performative outcomes so that every expression is nothing but the result of the particular functional configuration that the system gave to the endowment. The map defines—or, one could say, virtually subsumes—a set of possible paths without establishing which one should be taken. But then, a map is much more economical than a sequence of individual paths to manage the changes that may intervene in one's path and also to flexibly adapt the paths in ordinary situations.

On the other hand, what does it mean to shift from an "automatism-based model" to an "instrument-based model"? Well, if one considers the concept of ownership a prerequisite for the debate on subjectivity, then this shift means a lot. While the automatism governs the individual, the instrument *is used by* the individual. One could think of natural or acquired endowments as an instrument such as the hammer: I use the hammer; it does not control me. There is an important difference: while an automatism would transform the animal into a puppet moved by wires, the instrument is controlled by the animal who has broad flexibility in use. The functional completeness of the automatism would transform the individual in nothing more than the sum of its automatisms. This modelling metamorphosis, for example, is easily proven by the phenomenal analysis of the animal behaviour in the learning process: once an animal has learnt an operational schema, it then uses this schema in situations different from those in which it was acquired. Also, it uses it on the basis of flexibility and functional co-option, recalling Piaget's dialectic of assimilation/accommodation.[5]

The endowments can therefore be described as a "modal schema" that is body-referred (that is, it refers to the different functions of the body) and body-topical (that is, it identifies some specific involvements of the body). Furthermore, its structure implies and therefore allows for more functional paths, depending on the specific configurations of activation. The modal schema is therefore dynamic, multi-functional and evolutionary; its fractal structure in the synaptic networks allows the body to be relationally present in space. In other words, the endowments are the body's instruments and not vice versa. In order to offer the best performance to the relational needs of the body, which moves in an ever-changing and therefore unique reality, the endowments cannot give rise to static, mono-functional and unidirectional structures—like the stimulus-response interpretation would suggest—and must rather predict enactivism. In other words, the modal schema can neither be assimilated to a sensory-motor "unity of expression", just like in the behaviourist interpretation, nor can it be assimilated to an "amodal representation" of the external reality, just like in the cognitive interpretation.

To consider knowledge as the definition of a specific "modal scheme" implies that cognition does not fall outside of the body, it is not amodal or computational and it uses instruments that reproduce virtual fields of somatic relationship with the external reality. In this sense, a modal scheme is a sort of map of possible body-world conjugations related to a particular type of somatised dialogic states. The modal scheme defines some "useful functional range" so as to decline in a certain way the relationship between body and external reality. The body, therefore, indicates a field of relationships—this is the ultimate heterotrophic and vagile constitution of animality—declined through modal expressions. This relational state or embodiment cognition concerns a plurality of elements that cannot be translated into—or reduced to—the mere sensorimotor dialectics. Subjectivity, therefore, emerges from the systemics of this relational constitution that dialogues with the world through useful modal schemes.

The paradigmatic transformation implied by this modelling metamorphosis concerns different aspects: (a) the parameter of instrumentality or of the expression's distancing from the endowments themselves; (b) the parameter of modality or assimilation of the endowment instrument to the body; (c) the parameter of

[5]The child, for instance, uses processes of ex-aptation or assimilation-accommodation to build new representations, therefore the sympathetic elements give him the possibility to anchor the unknown elements and to extend the representational horizon of the child. The epiphany effect raised from the recognition in the non-human animal, the co-feeling in him, but at the same time in the being gradually projected in another dimension. S.J. Gould, and E.S. Vrba, *Exaptation-A Missing Term in the Science of Form*, «Paleobiology», Vol. 8, No. 1 (Winter, 1982), pp. 4–15, and J. Piaget, *Science of education and the psychology of the child*, New York: Orion Press, 1970.

relatedness or conjugative capabilities of the endowments. The instrument-based model gives an endowments-based character to the phylogenetic and ontogenetic instruction, in the sense of "being useful for"—that is, to structure possible "expressive modals" instead of identifying them, that is, instead of looking for "expressive motives". Subjectivity lies in the denial of this "functional completeness" that would turn the animal into a set of mechanisms activated by external or internal stimuli. By considering the information (natural and acquired) as instrument rather than automatism, one lets the animal be the owner of its endowments: it uses them; they do not control it. An instrument can be used in a thousand different ways, and with creativity as well. To be subjective means to be the owner of one's own instruments, to test them in new situations, to modify them and to create new ones as well.

But why is an instrument-based model more suitable than an automatism-based one? The answer lies in the "principle of singularity of the real"[6]: the world presents itself in similar but never identical conformations, so that the individual is always faced with some "margins of newness". If one's endowments were automatisms, one could not face the changes, because the automatism does nothing more than repeat the accredited function. For this reason, a software that plays chess through an algorithm would not be realistically possible. In order to control singularity, the animal must be able to use its endowments in a free and flexible way, through *ex-aptation* and with creativity. At this point, the evolving nature of these endowments becomes clear: they are constantly transformed according to the individual's specific use.

6.2 Subjectivity as Presence and Systemic Emergence Compared to the Contents

To connect subjectivity to a relational systemics of the body, whose multilayer of positionality exceeds the input-output dialectics, means to overcome Descartes' dualism that is still present in cognitivism, makes the individual's *Dasein* an abstraction and desomatizes subjectivity. The disjunction from the body may indeed be pursued not only by admitting a metaphysical *res cogitans*, but also by reading it through informatics and thus translating it in an amodal way, just like it happens in the computational formalism that tries to interpret the cognitive function. Reason,

[6]Knowledge anticipates being able to use a foundation of stability-familiarity to project oneself into an unknown dimension, which in turn can construct a new "plane of reality"—to quote Speusippus as recorded by Aristotle—and so to begin a new adventure. Leaning out into an unknown dimension remains dizzying, and one cannot make the transition without the assistance of an other: to construct this dialectic, the hybridizing process, which can open new epistemic dimensions, is indispensable. Cf. R. Marchesini, *Knowledge and Different Levels of Reality*, in «Reading Philosophy, Special Issue on Realism and Anti-Realism: New Perspectives», edited by L. Caffo, S. de Sanctis, V. Santarcangelo, N. 2, 2014, pp. 53–64.

language and intentionality, along the lines of the post-Cartesian philosophy, become disjunctive principles-operators compared to a *res-extensa* body that continues to be interpreted as a reactive entity rather than as a presence-relationship. This is a passive body that, affected by a stream of world, reacts in a mechanical way. According to dualism, one is a subject inasmuch one is emancipated from one's own corporeal root and, by assimilation, from one's own animality. In this paradigm, there is no room for an animal subjectivity. Moreover, the human subjectivity is separated from the meta-predicate of animality as well: one is a subject despite being an animal.

According to the mechanistic explanation of the *res extensa*, animality is not a relational positioning, that is, a *Dasein*. Rather, it is simply a reactive structure, that is, a set of motions activated by the stimulating and elicitative intervention on access-locks. If the animal-machine paradigm remains intact, some exhaustive models of expressiveness will inevitably be identified. Such models are able to fully explain the expression that the individual manifests in the timeless immediacy of his being. Martin Heidegger eloquently explained this subject-less animality as the stunned condition of beneficiary-reactive immediacy that, lacking presence, cannot be present.[7] However, this is a *petitio principii*: the model that explains the animal expression was intended to fully respond to the mechanistic requisites of the automaton-animal paradigm. There have been several debates on what kind of machine the animal is, but the paradigm that relates the animal condition to the machine has never been questioned.

Moreover, I believe that the reference to the machine is important because it highlights the great differences between the animal and the machine. First of all, it is necessary to say that, despite the apparent coherence with the scientific canon, the statement "the animal is a machine" does not meet the criteria of falsifiability of the hypothesis since no machine, even as a theoretical formulation, can do what even the simplest of animals can. However, the specific characteristics of a machine are the real problem:

(a) the completeness of its functional contents-directions compared to the performative outcomes;
(b) its self-containment, that is, the fact that it is never in a developmental condition, even when it introjects outside information;

[7]Heidegger clarifies Descartes' unsaid. The mechanical translation of animality created an unbridgeable gap between humans and other species and, simultaneously, it definitively sanctioned its operative freedom over the non-human universe. Since the seventeenth century, this paradigmatic development has been undoubtedly difficult and controversial: the point was to accentuate the dialectics of exclusion that was only in nuce in the early humanists. Therefore, it is no surprise that in addition to the hypothesis developed after Descartes' death (which increasingly amounted to a reductionism of animal expressiveness), proposals to restore subjectivity in nature came back, albeit in an intermittent way. On the other hand, countless factors privileged the reductionism to which Descartes offered a particularly effective paradigmatic crux—first of all, the autocratic and autopoietic operation of the human being as the sole protagonist, a principle that had already been active in the Western humanist metamorphosis for two centuries.

(c) its lack of only functional coordinates among the functional motives;
(d) its isochronic and therefore atemporal condition compared to the unavoidable diachronicity of a living organism;
(e) its lack of teleonomic structures relatable to the individual's past motives.

When one observes an animal and tries to interpret its behaviour, one is inevitably faced with an "explanatory deficit". Therefore, one usually adopts epistemological strategies based on causal plurality, such as the presence of both proximate and remote causes in Ernst Mayr's formulation or Niko Tinbergen's four questions of ethology.[8] The mechanistic translation cannot grasp the surplus of the animal behaviour; this can be grasped only by confusing the shared styles with the individual's actual expressive interpretation. Such a process can be attributed to an epistemic bias of categorization. Animality is thus the individual's unique position in the here-and-now: the individual's ability to interpret her own here-and-now with creativity, flexibility and partiality. One can therefore say that if the endowments, either natural or acquired, were a script for the individual to follow, the animal would be the result of the actor's interpretation rather than the mere repetition of the contents of the script.

If the endowments are considered as tools rather than automatisms, the expression cannot be derived directly from these; it should rather be derived from the positional systemics of the individual who bends the instrument to a particular function. The behaviour is thus a manifestation of the animals' unique and systemic state. The animal uses its own modal schemas by extracting the specific function from the range of virtualities required by the schema. Mentioning once again the example of the script, one can thus say that subjectivity lies in this interpretative singularity, in the actor's ownership of his part, on the basis of an overall status. In other words, the best way to understand subjectivity is to consider it as a result of an emerging singularity of the systemics. The animal expression, therefore, should not be regarded as the sum of the mechanical expressive units, but as an overall state of the system that, as a whole, extracts a particular functional declination from the functional virtuality of the endowments.

Subjectivity is thus the animal expression that cannot be reduced to its endowments, despite the modal value that they introject. When one talks about a

[8]These two statements are misleading, but they offer themselves to reifying projections. Starting from statement (a), both Ernst Mayr, when talking of remote and proximal causal duality, and Nikolaas Tinbergen, in posing his four questions, exhorted us not to conflate evolutionary compatibility, dictated by the fitness of the subject, and elicitative compatibility governed by hedonic-elicitative principles that the individual is subject to. So if it can be proven that the fitness of altruistic behavior responds to the parameters of Hamilton's coefficient of relatedness, then it is equally true that an individual possessing a given trait may express it toward any entity which is capable of eliciting it, for example the expression of maternal instinct by an adult female of one species toward a cub of a different species. The explanation that refers to fitness tells us only whether an attribute is compatible in individuals of a given species, not why an individual would express it. Cf. E. Mayr, *This is Biology: The Science of the Living World*. Cambridge: Belknap Press of Harvard University Press 1997, and N. Tinbergen, *The study of Instinct*, London and New York: Oxford University Press. 1951.

systemic condition and an emerging singularity, one refers to an entity that owes its virtuality to the redundancy—rather than linearity—of the expressive mechanism. The body is not made of strings and pulleys; it is a living entity that is constantly growing and converses with the outside world so much that every liminal definition is arbitrary. Where does the body begin? Where does it end? The answer would be apparently simple but, on closer inspection, every definition of "border" is arbitrary. The body is placed on a variety of levels (sensory, metabolic, endocrine, immunological, motoric, symbiotic) that converse with the outside world. Also, positionality—the emerging singularity—is the chaotic outcome of this continuous chatter of the body. In this sense, the subject's mental state can be considered not as a *homunculus*, but rather as a set of overlapping hierarchical levels whose predicates appear before the units that compose it.

Consequently, there is no control unit or place of subjectivity, but an emergence of functional states that define "different levels of subjectivity". These levels must carry out different functions, since the subject's behavioural needs are different. However, in their entirety, they define an inner world assembled in a way that is anything but logical, with very dystonic identity instances. Diachronically, the emerging positionality creates a continuity of identity: an inner world that, through its partiality, expresses this momentary "feeling a certain way"—the title of the singularity. To be subjective means to have an identity—a biography—that is the outcome of several moments (the history of the species, gestation, developmental age, adulthood) and relationships with the world. These stories coexist and each one bears specific motives. A causal plurality is therefore determined: the subject is always a Harlequin serving several masters. These stories are summarized in the individual that, precisely for this reason, is singular: this means to be unique and unrepeatable but, above all, it means that an animal's behaviour can never be predicted in detail.

However, this unpredictability should be attributed to being part of different causal temporalities rather than to the mere fact of being a complex system. The individual suffers from what her species has phylogenetically learned and introjected within the genetic and epigenetic inheritance of phenotypic translation, within the parental and cultural structures of ontogenetic translation. The individual is affected by both the first relationships she had with her parents as a baby and those that she had with her peers during childhood and that built her basic character. The individual is affected by her experiences and by the proximal plan of experience introjected. She has developed an evolutionary differential that depends on the received stimuli; she has built a specific knowledge. She responds to all this. Biographical identity is a work in progress, because the animal is constantly learning and therefore constantly changing its identity. The expression of a given behaviour is also the exercise of certain behavioural traits, so that every time that an animal performs an action, it transforms its own identity. The constant evolution of identity could be compared to muscles that are trained and therefore grow while working.

Therefore, to be subjective means to change, to never be equal to oneself, to be unfaithful to one's past. Also, learning is not a particular moment of subjectivity, but the very foundation of the subject: subjectivity means to possess an inner world in constant transformation. To be an animal means to dream, plan, build strategies, reflect on possibilities, evaluate the hidden contents of a situation, do mental simulations on different tactics. Subjectivity is the result of the animal's mental life, that is, a sort of inner theatre where each orientation is compared with all the others: Lorenz called it the "parliament of instincts".[9] The result is therefore always systemic, reflective and interactive involving different parts; it is never mechanical, analytical or like a ripple effect.

The mind as a system gives life to thought. It retreats into itself and leads to more or less conscious global decisions. However, the mind is not consciousness: consciousness is just one of the many functions of the mind. Moreover, mental subjectivity takes place especially in the unconscious: nothing is more subjective than the unconscious that manifests itself through desires, dreams, states of alteration, emotional images, unfolded memories, confused projects. In fact, one could say that consciousness is often a censor of subjectivity. The mind is an internal ecosystem: a microcosm that cannot overlap the macrocosm world in which processes of reality take place. The mind puts the animal being in the world, but at the same time it allows it to distance itself and reflect on given situations through a complex elaboration of all available information. The mind can transform all the events of the world in meanings—that is, it can translate them into categories of risks and opportunities and therefore perform an immediate decoding that allows it to make decisions.

6.3 Psychic Emergence and Positional-Relational State of the Body

The idea that the mind emerges through levels of appearance from all over the body and is not simply related to it, leads one to consider the psychic phenomenon as the result of being-a-body, disagreeing with Descartes' having-a-body and all its amodal and representational variants. The animal's ontological level must therefore be redefined and subjectivity must be brought back to the very character of animality. To be an animal does not mean to orient oneself towards something in an exclusively motor sense, but rather to assume a constitutive lack that demands to be revived through the relationship with the world. The body is shaken by positional states in its very metabolic entity that "is-done" and can never be defined statically. This means to distance oneself from the mechanical model that, at any time, can always be indexed in terms of functional contents from which to deduce performativity.

A metabolic entity constantly redefines its terms at stake. In other words, it is never executive because it always tends to overcome itself. This constitutive

[9]K. Lorenz, *On aggression*, New York, Harcourt, Brace & World, Inc., 1966.

circularity interrupts the subject-world disjunction as phenomenology emphasizes; I am referring in particular to Merleau-Ponty. When I talk about the unique emergency of positionality, therefore, I do not consider the problem of predictability as a distinctive meta-predicate of subjectivity: behavioural phenomena with a high degree of predictability can appear as, on the other hand, there are abiotic events whose course is absolutely unpredictable. I would like to point out the absolute arbitrariness of the explanatory coordinates of cause-effect. If the mind emerges from the entire body, it makes no sense to consider the mental state as the representation of something. In fact, it would be more plausible to somatically correlate the concept of intentionality: to be a body means to refer to something external—that is, to refer to something, to build one's own contents co-factorially.

The body can thus be regarded as a field where multiple games between different teams take place—phylogenetic instances that are related to ontogeny, needs that are confronted with environmental opportunities, stimuli that have to deal with the subject's here-and-now dispositions, memories that casually emerged during the dream phase changing the present prospective. Also, subjectivity is nothing but the set of results of this body-world conjugation. Subjectivity is an "expressive mode"—that is, a state declination. Therefore, it becomes unintelligible if one claims to extract it from the individual alchemically and in purity, by referring to the brain or some part of the CNS and failing to extend this ownership to the body. This is the most important reason why I do not think that an animal is a machine, however complex, unpredictable and interactive it may be: a machine performs functions *on* the environment, an animal achieves positional states *with* the environment. The concept of *res extensa* is based on the principle of the entity's completeness in the performative explanation, but the animal is not an entity that can be explained regardless of its positional circularities. On the other hand, it is clear that the principle of causal/functional completeness applied to the *res extensa* is inevitably reflected on the *res cogitans*. This is what leads Descartes to seek the certainty of one's own existence by retreating into oneself and questioning external reality.

Subjectivity is therefore a condition of positionality that arises from the animal's conjugative character. An animal realizes its subjectivity through its conjugative links to the world, by declining itself through connective structures. Every animal expression can be traced back to verbs of conjugation and positionality: chasing a prey, rejoicing during a game, being frightened by a danger, protecting a certain resource. The animal is brought into the world through a verbal predicate that is realized in the conjugative act, as if the world was to complete the sentence and give meaning to the verbal declination. The animal cannot be explained through an internal recognition (that is, *iuxta propria principia*) because it is implicitly (verbally) conjugative. Animality is thus a state of foundational deficiency, an appeal to something outside or past, and this will always distance it from the self-founding condition of a machine, regardless of the approximation or simulation that one may try to place in one's modelling: the phantom of the machine will always disappear.

The animal is subjective and, because of its desires, it is the protagonist of its own life. It is not a passive entity but rather a continuous self-presentation, an irrepressible being that is looking for something. There is a close relationship between animality and research, between subjectivity and non-equilibrium. The animal is driven by languor, libido, instinct. It creates situations, transforms the world into a field of opportunities, because it desires and thus looks for opportunities. A leaf is moved by the wind; the animal, instead, has an internal driving force: it is subjective because it is the protagonist. The internal motives are verbal structures: to collect, to chase, to look after, to join, to possess, to protect. The object is only an excuse. Desires are verbs, actions in power, "structures that connect"[10] in a certain way the animal to the world. The animal desires by expressing its motivations, by exploring and interacting with the world.

On the other hand, the motivational state is the direct and neuromodulated expression of the individual's endocrine status. For example, it expresses the individual's biorhythm and periodicity that, through internal rhythms, impose glandular secretions interacting with streams of light, food metabolites and immune feedbacks of the cytokine. Desires define behavioural motives that, compared to the individual's psychological state, are far from being amodal: one's thoughts are governed by verbal functions that, compared to the motivations, are perfectly modal. When one talks about self-assertion as an amodal function, one forgets that it is nothing more than a supervening level of basic motivational structures whose inductive and coordinative source are the gonads. When Heidegger says that the man is the child of care, he forgets that the epimeletic behaviour is based on oxytocin.

To be sure, I am not at all trying to reduce the behaviour of caring to the mere presence of oxytocin. I just want to emphasize how important it is to re-establish the connection between mental state and state of the body, avoiding the mind-body disjunction that only admits a correlation even in the domain diversity. Descartes' artifice of the dualistic operator opens the way for an amodal conception of cognition that is the first step to permanently remove subjectivity from the human and non-human animal condition. On the contrary, the expressive and relational performance of the mind always starts from a systemic-positional condition of the body: one only has to change some positional parameters and the psychic outcome assumes a totally different declination. Moreover, the thought always has a modal structure, compared to the body. On closer inspection, even the human language reflects the "propositional structure" already implicit in the positional act, rather than following formal rules. Chasing a hare is a behaviour or a thought, a statement or an implication: its internal structure is what really matters. This is defined by a motivational predicate (chasing) and a modal specification that comes from schemes of experience in relation to the environment (how, what, where, when). Chasing is a positional state because it defines the conjugative motive, creating an accurate declination of the connection between the subject and its here-and-now.

[10]G. Bateson, *Mind and Nature: A Necessary Unity* (Advances in Systems Theory, Complexity and the Human Sciences), Hampton Press 20002.

Positionality is given by two forms of conjugations: (1) proactive positionalities that indicate how the subject engages the external reality, incorporates it and orients itself therein; these positionalities can be defined as motives or "modal desires"; (2) reactive positionalities that indicate what kind of willingness the body has in the intersection with external reality; these can be defined as emotions or "modal willingness". Some authors such as Damasio have stressed the importance of these positionalities directly referring to the systemic body—and therefore defined as "somatic markers"—for cognitive activities usually defined as amodal. A significant example is the solution to a problem. First of all, it must be said that the animal does not just solve problems, as if external reality was nothing more than a series of objective problems. The motivational conjugation is the driving force that creates the problem—that is, it defines the gap between state and desired positionality.

Desire is the most immediate expression of subjectivity. It goes beyond need and, in a sense, gives meaning to it: I live because I desire, I live by expressing the desires that overwhelm me and shape my presence. All of the animal's endowments do nothing but offer themselves as instruments for expressing desires. Desire colours the world, it fills the eyes of a child's wonder, it supports the chaotic games of a puppy and gives its own meaning to the events of the world. If desire disappears, life withers into a vegetative atemporality; if desire decreases, life fades. Desire is a mental languor, a craving for the world that makes the individual restless, ready for action, creative and involved. Desire has always been at the basis of ethological research: each species has different desires, but every animal desires individually. It is a subject because it is oriented to seek a space where it can act: desire faces it with problems, makes it evaluate situations, urges it to find solutions, forces it to decide. The exercise of desire gives it more pleasure than the achieved outcome: the animal is a subject because it is part of the principle of expressive pleasure.

Secondly, it should be emphasized that, once a problem is revealed through the exercise of the desiring declination in conjugation with the outside world, this must be understood in its two inflections: (a) all the structural features of the problem must be accepted and the individual must therefore be able to refer to the positional resources that are useful for grasping complexity; for this reason, some problems require emotional activations of opening and others require emotional activations of closure; (b) the subject itself must be understood within the problem, as if the problem were a more or less comfortable environment, depending on the individual's intrinsic conditions. The animal is subjective because it has emotions, that is, some inner states that lead it to evaluate situations in a certain way. It is subjective because it does not experience the world passively but interprets external reality depending on its status. Through the emotional state, the animal converts the experienced situation in a value-for-itself condition. In this sense, to feel emotions is a necessary condition for a specific here-and-now: to be subjective means to build a singular and unique *Dasein* of both the individual and the momentary past.

Emotions are the result of the relationship between the condition of the body and what the external reality offers at a particular time. The body is a growing systemic reality: to transform this physiological processuality, which interacts with the world, in a state of mind means to exist as becoming. It is as if the animal coloured the world through its emotions: the world takes on different shades depending on the animal's feelings. For example, if it is tired and has negative feelings, it will not see what surrounds it the way it does when it is happy. Emotions such as sadness or fear will transform the world into something dark. Therefore, to be subjective means to have a partial reading of reality. The animal never reads what surrounds it objectively; its approach is subjective because partial and relating to its internal condition. Subjectivity is thus an interpretation of the world that anticipates the representation, a previous "evaluation" that transforms the individual in a pre-conceptual entity. The subject lives in the world but, more importantly, it lives according to its own values that colour the world in a certain way. The animal is partial; it always has preconceptions.

6.4 Subjectivity Means Existential Plurality

The ethological principle according to which the individual is immersed in the world in always-different ways can be traced back to Darwin's principle of adaptive evolution. However, this is neither an improvement nor an emancipation from a particular previous condition, as the evolution of the human being is often trivially taken to be. It is rather a specialization, that is, a special connection to an environment and lifestyle. Evolution is a dialogic path that takes place over time and in a context of plural relationships; it is the introjection of relationships that unfold between life and external reality and whose outcome gives rise to a virtual range of ontogenetic phenotypes: a morphoAutomata (animal) that is not only functional architecture but also a dimension of presence and expressiveness. To consider evolution as a dialogue means to distance oneself from:

(a) an autopoietic vision of the living being and the various revivals of Lamarck's idea that evolution has an overall or intrinsic purpose, that is, it pursues a project;
(b) the idea that the individual is a completely passive entity in the evolutionary process when, on the contrary, the action and the creativity that she expresses in her life act on the selective pressures and lead to a modelling shift.

Darwin's revolution must therefore be regarded as a work in progress, a workshop of debates on animality that leaves naturalism so as to become the paradigmatic crux for philosophical reflections on ontology. If one reads Darwin's thought correctly, one quickly realizes some important shifts from the humanistic tradition:

(1) the organ precedes the function and the anthropomorphic principle of being designed for a particular performative declination cannot therefore be applied;
(2) each organism is the result of a repeated modelling based on selective pressures, so that the living organism is an artefact and a work in progress;

(3) the structural emergence is not based on an ideational coordinate, there is no designer, and it abides by bottom-up (rather than top-down) organizational logics;
(4) there is no model or reference and the individual, therefore, can never approximate something; one cannot talk about gradients of completeness nor perfection, because evolution is a do-it-yourself process;
(5) the evolutionary process does not abide by deterministic coordinates and therefore cannot be compressed into an algorithm, because phylogenetics is historically based on occurrences and casualness.

These shifts are substantial and cannot be confined to naturalism. They rather enter the philosophical debate and the theme of animality leads to a comparison with the foundations of Western thought so that, in order to emphasize the discontinuity that Charles Darwin's thought gave life to, one has to talk more and more about a post-humanistic age, seriously questioning:

(a) universalism, that is, the presumption to identify a foundation of the projective and adaptive canon of humanism;
(b) essentialism, that is, the idea that each entity needs to follow a previous authentic, unchangeable and referential schema;
(c) dichotomous thought in its various forms—particularly those of nature/culture, mind/body, human/animal, natural/artificial, innate/learned—that is, the idea that juxtaposed or related dimensions have a different domain;
(d) the disjunctive or non-relational conception of entities, that is, the idea that an entity may be founded according to its own principles and present itself as an impenetrable and autopoietic entity that can be reached through internal recognition and reflective speculation;
(e) the anthropocentric view of the ontological constitution, that is, the idea that the human being is measure and subsumption of the world.

Plurality entered the ontological debate with von Uexküll's concept of *Umwelt*,[11] that is, the immersion of a specific species in the world. This interpretation was then taken up by Heidegger so as to limit animality to the heterospecific and, vice versa, so as to free the human being from a particular declination, following to the letter the dualistic tradition of the Epimethean genealogy of non-humans as opposed to the Promethean genealogy of humans. By eliminating or drastically reducing the Epimethean features of the human being—see the theory of human incompleteness that goes from Pico della Mirandola to Arnold Gehlen—one tries to find congruity between Darwin's thought and the humanistic paradigm. However, by doing so, the human being's basic premises are misdirected. The animal, confined to its immersion, does not interpret its *Dasein* but simply enjoys and responds directly to the stimuli of the world in an absolutely atemporal stunned

[11] J.J. von Uexküll, "A Stroll Through the Worlds of Animals and Men. A Picture Book of Invisible Worlds". *Instinctive Behavior. The Development of a Modern Concept*. New York: International Universities Press, 1954, 5–80.

condition. Once again, one tries to desperately save the human's speciality that guarantees universalism.

According to this monadic vision of the immersion, the non-human animal is an other that cannot be investigated—as explained in Thomas Nagel's famous essay[12]—that is, an entity with which no empathic relation can be built since there is no connective bridge. One is once again faced with an ad hoc interpretation that expunges animality from the human domain and reaffirms the automaton model. Actually, if one reads evolutionism correctly, one notices an extensive overlap between different *Umwelten*. This is attributable to:

(1) homologies, that is, characteristics that are antecedent, shared because inherited from a common ancestor;
(2) analogies, that is, characteristics that are convergent, shared because produced by the same selective pressures. Diversity, therefore, should be not be used:
    (a) to establish a total diversity between non-human and human that justifies an ontological difference;
    (b) to build an existential hierarchy, make the human an evaluation parameter for the non-human and the latter an approximation;
    (c) to declare a total alienation of the non-human and therefore the impossibility for any hypothesis on heterospecific existentialism that is not a mere behavioural description.

Subjectivity thus enters the somatic-phylogenetic dimension as a body's specific modality. Human subjectivity, therefore, is not the result of emancipation from a generic animal condition, but rather the very expression of a specific animal condition. Subjectivity is the systemic emergence of a body in dialogue with the world, a being-with that subsumes the previous dialogic instances-consequences by constructing a new relational field that "exists" insofar as "it-is-done"—that is to say, the two states do not differ but rather collide. Existential plurality has several levels of contingency:

(a) the phylogenetic particularity that does not only specify absolute distances (human greater closeness to chimpanzees than flies) but also some shared features that vary from species to species;
(b) the ontogenetic state that is not only an individuality that can be defined depending on a cofactorial multiplicity (genetic, epigenetic, experiential) but is also the presence in a certain age;
(c) the singularity of the here-and-now, that is, the state of the body and its relationship with the events of a particular time. Subjectivity lies in this stream-like and continuous oscillation of presence in the world.

The humanistic paradigm tends to expunge diversity so as to realize its project of universalism, of formal and expressive plasticity. The non-human diversity is

[12]T. Nagel, "What Is it Like to Be a Bat?".

therefore deprived of subjectivity and turned into a mechanic specificity; it is transformed from existential diversity to functional diversity. For this reason, it seems very difficult to acknowledge a plurality in the *intus-legere* process, transforming it in a multiplicity of reactive mechanisms. Twentieth-century cognitive ethology has not answered by questioning ontological anthropocentrism—that is, the cognitive universal—but rather by presuming that the other species participate in the same *intus-legere*. In my opinion, this is the greatest paradox that has removed rather than introduced a debate on animal subjectivity. Just as a morphological, endocrine and sensory diversity exists, so does a cognitive plurality, since the phylogenetic process specializes but does not improve. The animal world is thus formed by a variety of intelligences specialized to process present information. Therefore, to ask oneself what is the most intelligent animal makes no more sense than to ask what is the most sensory or endocrine species. A dog, for example, has a strong social capacity. A cat, instead, can solve problems independently: to ask which species is more intelligent is absolutely useless and wrong.

To acknowledge subjectivity in the animal condition—that is, to bring subjectivity back to its animal condition—means to open a debate on many of the philosophical premises that characterize Western culture. These premises tried to solve the problem of subjectivity through a series of artifices: the most notable one is the dualistic artifice of the incorporeal or amodal presence in time. Subjectivity cannot be extracted from the body and cannot even be confined to a particular emergence of the evolutionary plurality, neither taxonomically (to what animal can subjectivity be related) nor phylogenetically (from what point of the human evolution onward one can speak of subjectivity). A living being is a subject because its ownership emerges from its systemic dialogic presence, from its reference to something external but already presumed (like a lock that requires a certain key) and in progress (like a performance that always modifies the script). Subjectivity indicates the individual's escape from the already-given: the excess of his being.

The animal creates its present time, a here-and-now in which it builds its *Dasein* (being-there). This present does not exist in itself and the animal invents it by relating past and future—that is, by giving resonance to the passed time and thus creating a temporal space for action and presence. The animal is a subject because it exists: it does not simply appear in the world; it is present, that is, it participates in the world. To own a present means to have sovereignty over the here-and-now and, in this sense, not to be immersed in it, but rather emerge from the fluxes of time. The animal expresses itself through codes of resonance of the here-and-now that change depending on the species. One could therefore say that there are as many presents as animal species. Each species relates different times because the phylogenetic history that preceded the individual emergence of a species-specific subjectivity is different as well. Subjectivity is this emergence from the flux of time.

One does not choose to be a dog, a dolphin, a man or a hen. One does not choose to bear the legacy defining the particular characteristics of one's subjectivity. One does not choose the ontogenetic path that defines this individual identity, of which one so wishes to be the undisputed and sole creator. One does not choose the here-and-now that flexes one's body unpredictably and ungovernably, like a slender boat trying to stay on course amidst stormy waves. Although one did not choose any of these states and legacies, one does not give up; no individual does it, whatever species one had the fortune to belong to. The idea of the animal as a machine is finally declining and it is now necessary to reconstruct an animal ontology that goes beyond the concept of *res extensa.*

6.5 The Emancipation of Animality

Life is about relationships developing over time, synchronized and incomprehensible causalities that constantly demand remote explanations of the text. Life shows its corpuscularity in the here-and-now and yet also expresses an undulatory nature, a choral Dionysian flow deriding any explanatory attempt. Life is a succession of thresholds connecting seemingly distant worlds, in the ambivalence of processes that seem closed in on themselves and open systems, poetics that arise as fractalic references to an otherness that is constantly redefining itself. Life is a relationship, an itinerary drawn by heteronomy, a creative process that uses dialogue to address the uniqueness of the real.

One's presence in the world is a continuous process of mirroring the other, a being-with that implies a somatic communion, an ecstasy within otherness, a shared nature of bodies harmoniously aligned on the same level of relationship. Otherness gives itself through an affective process that is first of all an impression, a motion of co-sentience (rather than appearance or noematic content) and, simultaneously, an epiphany—that is, a revelation of existential dimensions virtually accessible for the self, a threshold to escape from one's individual boundaries, a suspension of such boundaries and a participation in something that transcends the self. Thresholds are never anodyne holes along the bastions of identity, but rather interfaces, filters that allow for the mirroring process. It is therefore important to consider ecological thinking not as a description of the living, but rather as a paradigm that interprets the living and its poietic skills. Life is creativity, as somatization of relationships and transformation of dialogues into flesh. Therefore, a threshold is the place of perception as experience and visitation of otherness. Life is not made of separate entities, but rather of connecting thresholds. Life is a continuous seminar of relationships. It is a flow that, just like a flame in need of oxygen, goes upwards and produces ever-changing figures that—unable to repeat themselves—throb in their singularity and take space through the hybridization with the world.

The interpretation of animal otherness is paradigmatic for the understanding of those mechanisms of removal and reduction that do not allow one to deal with the more general issue of hybridization. Typically, the non-human is subjected to the following nullifications of its potential presence to the dialogue:

(1) reification, that is, the negation of its subjectivity, which may find several forms, the most common being lack of will, non-possession of logos, instinctive irrationality, inability to achieve high levels of intentionality;

(2) transformation into a regressive figure, believing the animal to represent the past and the domain of animality to represent the heritage; this idea must be rejected in order to fully appeal to the human dimension. This denies the non-human animal an effective presence to the encounter;
(3) alienation, that is, the prejudice that the existential level of other species is unknowable, with the consequent inability to effectively communicate and establish a relationship, believing the different *umwelten* to be separate monads;
(4) projection, that is, the transformation of the non-human animal into a mirror that therefore does not mitigate but rather strengthens the subject's narcissistic tendencies, with the consequent arbitrary construction of the other's identity and positionality, which is in fact denied effective participation to any (even basic) form of the relational encounter.

Animal otherness, subjected to these operations of denial and removal, is distanced both from the relationship and from revelation. The threshold with the animal source, designated to open up to recognition (the awareness of animal-being) and epiphany (one's reflection in heterospecific otherness) therefore undergoes a final rejection, and what would normally allow one to participate in animality is thus removed. When animal otherness undergoes these forms of humiliation and rejection, part of the reflection always collapses. First of all, the human being tends to deny both the presence and the substance of its animal-being: (a) animality thus does not become a dimension allowing for a specific declination of the human, but rather the shore to leave behind so as to let the human emerge; (b) the animal drive is translated into heritage, projected into the past and considered a sort of ancestrality that rises from the depths and muddies the waters presence, rather than the reason for Dasein.

At this point, a redefinition of the character of animality is necessary. One has to abandon the reclusion inside the res extensa—the animal as automaton—that becomes a monadic bubble in the *Umwelt* and poverty-in-world in Heidegger's thought. If animality is deprived of any character of subjectivity and turned into a mere mechanical functionality of some automatisms, trained by phylogenesis and ontogenesis, then it is clear that any predication of the human will seek comfort within an emancipation from animality and not within the expression of its animality. This is why I speak of "emancipation of animality" from the captivity in which it was kept by Descartes. This is why, in the field of philosophical ethology, I call into question the meta-predicative structure of animal-being, that is, animal ontology. Animality is a creative, and therefore relational, presence in the world that takes place through the diachronic construction of states of singularity.

Animal-being thus means constantly transcending the past and to rearrange heritage according to coordinates given by the here-and-now. In this regard, the animal is not immersed in the present but rather builds its own present arbitrarily. For this reason, I speak of animal Dasein: if the animal were an automaton, she would freeze at the first change! However, given the singularity of the real—it offers similar, but not identical, occurrences—the animal has to be present and able to interpret the "margin of originality" provided by its here-and-now. The individual is an actor playing a certain script. Therefore, animal-being means "conjugating reality according to specific coordinates" or, in other words, establishing a unique dialogue with the world. This dialogue, however, should not be compared to a mechanical function—even if it is instructed by specific endowments, be it innate or learned. Endowment as such cannot bring out interpretation, which, on the contrary, is the core skill to manage the "margin of originality". Endowment is a set of tools, not of automatisms.

Animal subjectivity—which thus also defines human subjectivity—is a modal expression of body, as Merleau-Ponty had well understood, and as we can see in Francisco Varela's enactive conception. It can be linked to the condition of "being a body" that shows a plurality of relational levels expressing themselves through internal organizational dynamics and specific perceptive inclinations actively addressed to external reality. So, it is possible to say that subjectivity is not an amodal condition but an intentional structure requiring external specification. It means recognizing Dasein within animality. On the other hand, if it is true that every specific animal taxonomy indicates a specific immersion in the world, then it is also true that the *umwelten* have a significant degree of homological and analogical overlapping. Therefore, one is never facing separate worlds, but rather universes constantly dialoguing with each other.

Chapter 7
The Posthuman Dimension

7.1 Premise

I must explain the difference between the post-human view and the development of post-humanist philosophy, which is more interested in defining a connection with, and an overcoming of, humanism rather than defining the human as a predicative virtuality. According to post-humanist philosophical criticism, the term of comparison here is humanism—that is, a particular interpretation of the human and therefore an anthropocentric direction of ontology, epistemology, aesthetics, and ethics. Post-humanist philosophy therefore focuses on the meaning of *techne*, ontopoiesis, the relationships with otherness and with nature as a whole. Moreover, post-humanist thought focuses on the emergent meaning of relations, under the influence of authors like Gregory Bateson.[1] One would be tempted to recognize a kind of basic ecocentrism in this philosophy, if post-humanism did not reject any permanent axes around which to revolve. On the other hand, compared to traditional ecocentric theories—see Arne Naess' thought[2]—there is an obvious interpretative shift of the meaning *techne* and a manifest rejection of the "natural *versus* artificial" dualism that characterized the previous approaches.

7.2 Post-humanism *Versus* Trans-humanism

In order to understand the post-humanist perspective, one must go back to the post-human view and, above all, try to compare it with the trans-humanistic proposal that is generally more known because of its resounding and visionary power.

---

[1]This theory owns a lot to the ecocentric thought and complexity and, in particular, to authors such as Gregory Bateson, Michail Bachtin, Tzvetan Todorov. For Bakhtin, *Dostoevskij. Poetica e stilistica*, Einaudi, Torino 2002; and for Todorov, *The conquest of America*.
[2]A. Naess, *Ecology, community and lifestyle: outline of an ecosophy*, Cambridge: Cambridge University Press, 1989.

R. Marchesini, *Over the Human*, Numanities - Arts and Humanities in Progress 4, DOI 10.1007/978-3-319-62581-2_7

The trans-humanist perspective starts from an evident goal that could be summarized as the technological enhancement of the human being. One could therefore say that *techne* has a soteriological task, that is, it must save the human being from damnation and captivity, which can be identified in its finitude and phylogenetic declination (that is, within a specific range of performative potentialities). It is easy to see the dualistic vision of transhumanism, based on a revival of the Cartesian conditions of "owning a body", "being confined within a body", "being related to a body". The difference is that this perspective adopts an immanent res informatica instead of a transcendent res-cogitans, the former being a pack of information that establishes an amodal and therefore potentially undoable relationship with the body.

The mind-uploading project, for instance, rests on the idea/goal of being able to extract the mind from the brain so that it can be transferred to any other medium, and is one of the cornerstones of the transhumanist proposal. Such a project is based on a mind-brain dualism or on an amodal relationship between the two terms, as if the brain were a kind of hardware and the mind were software contained therein. The expected separation is possible because the mental software, defined in terms of content, can be implicitly extracted. This is so close to the concept *of res-cogitans* that the explanatory model could be said to be the same: the only difference consists in transforming the *res-cogitans* (counter-term of the corporeal *res-extensa*) from an unmeasurable and transcendent entity to an immanent reality—that is, *a res-extensa* that is incorporeal but measurable as made up of information. The mind, as such, revolves around the brain-hardware but does not depend on it.

According to this interpretation, the subjective identity—understood as a set of mental contents, be they memories, projections, inclinations, dreams or goals—is therefore said to be part of an "amodal relation with the body" rather than a direct expression of the body. This is why the individual's mind can be extracted and therefore saved from the decay of the body. To extract the mind from the body means to give it immortality without having to rely on an extra worldly dimension, but it also means to think of inhabiting different bodies (not necessarily organic) and of overcoming the phylogenetic dimensionality. The mind-information can therefore be placed in different hosting matrixes: a computer, a network, another body, a clone, and so on. Considering death as a curable disease, transhumanists therefore propose hibernation and support the micro-scansions of the synaptic network.

From the transhumanist perspective, technology becomes the demiurge able to introduce the human being to different worlds and contain her (obviously, this does not mean limiting her but rather defining her) within an amniotic fluid space formed by the technosphere. In this sense, transhumanism thinks that the human being is projected into the future. A linear and technological definition of the diachronic human being thus results when the post-human condition is a goal, that is, a horizon made possible by the exponential sequence of technological achievements. Technopoiesis is no longer an exclusively emanative-autarkic process of the human being, since the machine is believed to evolve and would therefore become a co-factor in the emergence of the post-human. Hence the clear difference with the anthropoplastic interpretation of technopoiesis as it appears in the humanistic tradition. However, this human projection toward an ascendent dimension—one that diverges from the phylogenetic condition and the co-belonging to animality—closely resembles Pico's interpretation and Nietzsche's overhumanity. Transhumanist imagery, despite its technophilic suggestion and vision, evokes a hyperhumanist interpretation.

On the other hand, transhumanist hyperhumanism is very different from Martin Heidegger's ontological hyperhumanism. In fact, Heidegger realized that after Darwin's revolution, traditional humanism could hardly give the human being a connotation of specialty—instead of just specificity. He therefore adopted the ontological distinction between human beings and the other species by means of the dichotomy poor-in-world versus creator-of-worlds. However, Heidegger suggests interpreting *techne* as "unveiling". As we shall see, this interpretation is closer to the post-humanist view than to the transhumanist interpretation of *techne* as "realization-telos". On the other hand, if one takes into account Nick Bostrom's, Ray Kurzweil's and in a way Peter Sloterdijk's considerations, one finds Nietzschean themes such as anthropotechnics and the ontopoietic ascension in the overhuman. According to the transhumanist conception, therefore, the post-human is a horizon that emerges spontaneously, whenever the technopoietic convergence and exponentiality give rise to what scholars such as Vernor Vinge called singularity—that is, a veritable techno-anthropological quantum jump able to break linearity in human development.

The post-humanist perspective is therefore very different from transhumanism, for several reasons:

(a) it derives from deconstructionist and postmodern philosophy, as it assimilated its criticism of the models and its critical view of *techne*, starting from the Frankfurt school until the great French philosophers of the twentieth century;
(b) it considers otherness as co-determinant and referential and therefore considers hybridization as the emergence of prospective plurality rather than as human enhancement;
(c) it is based on a relational conception of predicative emergence—that is, when the results are considered supervening and not implicit in the hybridization—and therefore refuses to assimilate otherness;
(d) it believes that finitude is a human value and driving force rather than a deficit to overcome or a gap to fill. It also believes that the technopoietic act is not a helpful tool but a generator of human deficiencies and needs;
(e) it considers *techne* a creative-dialogical act that, as an epiphanic/revelatory event (instead of amniotic/disjunctive one), strengthens the subject-world connection in a way similar to Heidegger's concept of unveiling;
(f) it considers subjectivity a modal and enactive expression—being a body—that therefore does not allow one to free or extract oneself from corporeity. Instead, it believes that one should see the foundation of Dasein in its somatic intentionality;

(g) it considers the human being's diachronic feature as a multi-circularity in which different temporalities are simultaneously present and do not only recur in the individual's subjective life but also in the human being's projections of her relationship with the world;
(h) it does not consider the post-human condition as a horizon or goal, but as the actual and necessary human condition whose predicates cannot be founded *iutxa propria principia* and are rather the result of a constant hybridization with external reality—that is, one can only be post-human.

Starting from the necessary character of the hybrid condition—which is a dialogical ontology compared to the reflective ontology of the humanist cogito—one realises that the post-human is not what follows the human but rather the set of all predications that the human being is able to express. However, since the core of philosophical reflection is not what is biologically human but rather the human's ontological characters, it would be a mistake to try to understand the human predicates starting from an enucleative analysis of the human being. This is why there is post-humanist philosophy or a philosophical post-humanism.

The post-humanist proposal definitely represents a deep break in philosophical thought. However, it would be wrong to consider it a form of anti-humanism, because this would mean redefining the humanist canon on the basis of some shifts that concern:

1. the ontological definition of the human condition;
2. the interpretation of animality;
3. the referential or co-factorial meaning of otherness;
4. the anthropo-poietic meaning of *techne*.

If one wanted to summarize these shifts in a "paradigmatic crux", one could say that from a post-humanist point of view, human predicates are neither based on the human being nor founded *iuxta propria principia*. This means that they are not based on an autarkic, autopoietic and essentially reflective conception of ontology. The human being is considered the result of an introjecting/referential relationship with otherness. Therefore, the predicate derives from the point of connection with them and is a dialogical result of it. In other words, the predicates are considered emergent rather than emanative. From a post-humanist point of view, the human being does not emerge from a disjunction or purification from otherness—that is, from a self-referential humanity (as in Descartes' cogito)—but, on the contrary, from a hospitable openness toward otherness. The human being, therefore, results from hybridization, an anthropo-decentrative event, an epiphany that opens up new perspectives for Dasein. Otherness thus activates existential dimensions that do not pre-exist the dialogical-conjugative act. One can say that the human condition is realized at the edge and through marginalization. Mostly, it supervenes the human being.

When speaking of a "referential meaning" of otherness, I do not mean the three classical conceptions of dialectics:

1. the *polemos* or comparison that defines predication through opposition;
2. the *mimesis*, that is, the assimilation or reproduction of the predicative characters of otherness;
3. the *taxon*, that is, predication as the relational position of otherness.

Rather, I am referring to an inclusive or organizational dialectics, that is:

(a) the human being's representation of otherness;
(b) the projection of the human being in otherness;
(c) the reorganization of the human through otherness;
(d) the human being's emergence as the outcome of this encounter.

This involves a series of metamorphoses that allow one to talk of a post-humanist approach. I would like to mention a few of them:

1. the overcoming of essentialism, that is, the idea that the entity is contained within and gravitates toward a model;
2. the refusal of the dichotomous-disjunctive model that defines the entity by creating impermeable domains;
3. the overcoming of the exclusive dialectics based on a universal concept of belongingness;
4. the virtual vision of being, which does not mean cancelling the internal predicates, but reiterating the importance of the organization operated from the outside;
5. the relational interpretation of the predicative outcome that should not be assigned to the mere characters of the related entities, but should rather be considered emergent compared to them.

Ontology results from relation, that is, a process that is neither intrinsic to the subject nor already given. The human being thus stems from the ability to interrupt a certain immersion in the real and open intersections and connections to the world. The human being does not manifest itself through the fractalic anthropoplastic reproduction that claims to measure and subsume the real, but rather by introducing new levels of reality through the hybridization with external beings. In this sense, one can say that the human being should never be considered defined or definable, but rather a constantly renegotiable position destined to overcome itself by its own existence. One could say that the human being cannot decide not to participate. However, this does not mean forgetting the core of the human condition, transforming it into an anodyne and totally empty liquidity to which any predication can be assigned, as if otherness were the container and the human being an amorphous content. What makes hybridization possible is the human being's specific phylogenetic legacy: a certain structure or nature. Which is virtual and non-deterministic. Therefore, this means that one should start from the human being's animality and recognize its predicative redundancies (which are not lacks, but rather the contrary) that allow for the heteronomic organizational plurality.

7.3 A New Culture for *Techne*

One tends to think that the human being uses *techne* in order to emancipate himself from the animal condition and withdraw into an amniotic universe where everything is under his strict control. This thought leads one to believe that the human being uses techniques and technologies in order to dominate the world and separate himself from otherness. This autarkic and autopoietic vision comes from the humanist tradition and, in my opinion, is today's most serious obstacle to the development of a broader interpretation of technology and, in particular, of the new developing technologies. The humanistic tradition, referring to the myth of Prometheus and Epimetheus, considered *techne* as a compensatory act, that is, a "crutch" that man uses to compensate for the lack of endowments that nature gave him. Obviously, this interpretation is retrospective: it is clear that a man surrounded by technical support perceives himself as weak and helpless if deprived of his instruments. However, the compensatory reasoning does not hold, for many reasons. I would like to point out two of them:

1. if technopoiesis were compensatory, it would have to decelerate instead of growing exponentially;
2. if one analyses the somatic and ethological characteristics of the human being as a species, one will not find any deficit.

The humanistic tradition considers the cultural dimension as complementary and therefore believes that the relationship between nature and culture is proportionally inverse. From a neurobiological point of view, however, this is a huge mistake. Ontogenesis is not a replenishment process for which the emptier the vase, the more one can pour into it. Ontogenesis, in fact, organizes the components and the evolutionary differential between the parts. Therefore, the ratio between innate endowment and learning potential is directly proportional. The system allows for a wide range of possible identity trajectories because it is redundant—not deficient. Furthermore, the relationship is dimensional, and not juxtapositional. The humanistic tradition considers *techne* an instrument to emancipate mankind from animality—that is, to create a clear disjunction between man and animal. The human being is seen as a tightrope walker oscillating between the animal condition and a condition of total emancipation from it (angel or superman): *techne* is his rope. This interpretation affects any human progressive event, so that even phylogeny is interpreted as emancipation from the original animal condition. This means misunderstanding Darwin as well as the effects of the partnership with the instrument: the evolution and the hybridization with the instrument determine specializations and therefore a greater connection to the substrate.

The humanistic tradition considers technopoiesis as an event of autarkic creativity, as if every technique and technology came directly out of man's mind, as Athena came out of Jupiter's. This emanative vision of *techne* strengthens the separation between man and the world, up to considering the human being impermeable to the influences of the non-human. The autarkic premise, in turn, is at the basis of the autopoietic conception of the human being: if man establishes and determines himself can be founded *iuxta propria principia* he needs no further co-factors to explain his connotations. Actually, though, technopoiesis always results from the relations with the non-human universe and is never solipsistic. This does not mean that *techne* is a mere act of imitation, but rather a re-interpretative or representational process: it is like an actor who plays the part subjectively. Technopoiesis emerges when whatever is external to the relation shifts from a phenomenal condition (other-than-itself) to an epiphanic condition (other-in-itself) That is, for example, when the flight of a bird makes the observer fly too, creating in her a new existential dimension that she has to reach through a technopoietic act.

The humanistic tradition considers the human being as an essence that can be recovered by purging him from the contaminants. The autopoietic principle thus becomes a rite of purification and *techne* is the officiator of this ritual that aims to detoxify the human being from the dirt of the earth and animality. On closer inspection, technopoiesis increases the human hybridization by introjecting the non-human existential dimensions achieved through techno-mediation. Every technology increases our contamination with the world and distances us from the anthropocentric gravitation. In this regard, it is necessary to consider another aspect of the humanist tradition: the exonerative principle. *Techne* does not exonerate us from the performance, but simply introduces new performativities that are primarily based on the ability to adapt to the instrument, that is, to realize the hybrid function. *Techne* does not distance us from the stimuli but rather introduces new interfaces and therefore makes us more sensitive to the stimuli. *Techne* does not lower the directional-instinctual meaning but rather ignites one's desires even more, by lowering the fulfilment potential of one's actions, stimulating them with easy gratification, and increasing the languor by decreasing the surrogates.

According to the humanistic tradition, the human being is decentralized from the world and this leads him to resolve his status of non-equilibrium by seeking stability in technical support. But it is exactly the opposite. Technology is like love, and instability is always ex-post: one feels unstable without one's beloved only after falling in love. Technology is what brings decentralization about—not the *ab-origine* condition that produces technology. Every technopoiesis distances the human being from her gravitational centre, making her increasingly unstable and decentralized. This is what transforms technopoiesis in a self-implementing and therefore accelerative event. Technology increases one's non-equilibrium, one's need to look outside, one's sense of lack. Technology is defined as amnion, comfort, and reorganization, and the human being is therefore believed to gain some sort of compensation from *techne*. Actually, *techne* dissects the body as well as the identity, creating different systemic involvements. Every technological emergence invents new spaces for somatic experimentation and allows the body to reinvent itself. The technical element melts the organs and allows the tissues to reach new conditions of systemic virtuality. In other words, every technopoiesis introduces performativities that are not present in the phylogenetic endowment, not because it adds something or exonerates from a function, but because it creates new connections within the body.

This allows me to address a new topic, that is, the humanist idea that *techne* strengthens the human predicates without questioning them. This tradition considers techno-mediation as a probiotic function, just as that by which intestinal bacteria that facilitate the enteric process. In fact, *techne* is like a virus that enters the cell and reprograms its function. *Techne*, in other words, gives new characteristics to the human being. However, *techne* does more than that: it intervenes on the entire human ontology by metamorphosing the predicates but also by introducing new purposes. And so one realizes that the humanistic interpretation gives *techne* an ancillary role: it serves human beings, is unable to formulate strategies and must simply obey the human, who remains firmly in control of his own destiny. Even in this case, post-humanist philosophy reads this relationship very differently. First of all, one can say that since the second industrial revolution, the relationship between humans and technology has become equal. However, with the dimensional transformation of technology (technosphere), its managerial character has become clearer and clearer. Human being serves technology—not the other way round.

The humanist tradition, in its ontological anthropocentrism, mistakenly considers technology as a glove that: (1) helps to compensate for the vulnerability of the hand; (2) enhances its capabilities without questioning it; (3) faithfully reproduces its anatomy; (4) serves it in whatever it does; (5) prevents contamination with external elements and therefore preserves its purity; (6) simplifies tasks, that is, exonerates it from some particular functions; (7) separates it from the world, that is, makes it impermeable; (8) embraces it creating a new micro universe that is a projection of it. In all these expressions, technology is juxtaposed to the body, is an emanation of the body and serves the body. Humanism has an ergonomic conception of technology because it believes that technology serves the body and not the other way round—that is, the body serves technology. But if technopoiesis is the result of an epiphany emerged through the connection with the non-human, it is still the body that serves technology and not vice versa. Humanism focuses on man's body—as an example of beauty, measure, subsumption—creating a plastic-anthropological conception of creativity in all its disciplines, including architecture and technology. The post-humanist interpretation decentralises it, not by choosing a different centre but by paying attention to the interfaces. The body therefore becomes a liminal film that can be interconnected—that is, a welcoming place.

From a humanistic point of view, the "Vitruvian universal" is also the exemplary image in technopoiesis, so that technology as an emanation becomes nothing but a fractal and stigmergic projection of the human body. When one speaks of an ergonomic, compensatory, juxtaposititional, ancillary, probiotic, disclaiming interpretation, one always refers to the Vitruvian universal coordinating technopoiesis. From a post-humanist point of view, the "cyberganic pluriverse" is the best way to understand the hybridising meaning of *techne*. When I speak of hybridization, I am giving a dissective, redundant, infiltrative, managerial, viral, conjugative interpretation and am referring to margins of interface and the dissolution of any centre. The hybrid accepts its dialectic nature. Obviously, this inclusive dialectic is very different from the exclusive dialectic of Heraclitus's polemos. The hybrid accepts that its body will host technological support, and therefore accepts the dissective logic that exceeds the ergonomic conception. The hybrid considers it a process that widens the spaces of virtuality and redundancy and does not compensate for any somatic deficiencies. The hybrid lets technology infiltrate it and does not claim to see it as a glove. It knows that once technology enters, it will not be an obedient servant but a manager imposing governance; it will not enhance the functions but subvert them like a virus; it will not exonerate but conjugate.

Post-humanist philosophy proposes itself as a new culture of technology and interprets it as a viral and infiltrative element that tends to bend the body toward new coordinates, producing instances and purposes that retrospectively emerge from hybridisation and introducing new *umwelten* and situations of adaptive non-equilibrium. Humanism thinks that technopoiesis results from rationality. Post-humanism, instead, believes that technopoiesis is mainly hallucination, psychedelia, possession, ecstasy—it means overcoming the stability and equilibrium of/in reason so as to project oneself in overhuman territories, achieving transpecific exchanges, increasing one's conjugative interfaces, establishing new dialogues with what is real and bringing out new levels of reality. Post-humanist philosophy considers technopoiesis as an anthropo-decentrative process. Post-humanist thought interprets the technopoietic emergence and its effects, starting from the concept of hybridization, in a new way. This means emphasizing the conjunctive and participatory role played by *techne*, its implementation of needs and conjugation with otherness, the induction of ontological and epistemological metamorphoses that invest the human being's interface as well as its identity coordinates and overall being-in-the-world. The hybrid thus becomes the clearest, most exemplary and complete expression of the overcoming of ontological anthropocentrism, which post-humanist philosophy stands for.

7.4 What Are the Ontological Differences Between the Vitruvian Model and the Cyborg?

1. the former places itself at the centre, it is centripetal; the latter focuses on the margins, avoiding any definition of centre, it is present in its connecting structures;
2. the former enhances its being an essence, therefore its purity, the latter is based on a principle of contamination, on its being a hybrid;
3. the former replicates itself in the world, it is anthropoplastic; the latter hosts the world, its boundaries are hospitable thresholds;
4. the former considers technology as an instrument of domination, it transforms the world into a laden table; the latter believes it to be a medium to connect to reality understood as a partner;

5. the former regards itself as a technical manufacturer, the Promethean human forging the world; the latter considers itself as a product of technology, Epimetheus's human son;
6. the former believes to be self-sufficient, autopoietic, and thinkable *iuxta propria principia*; the latter believes to result from the encounter with otherness and to be unthinkable through an internal analysis.

The general idea that results from the hybrid, as a point of ontological convergence and divergence, is that of a being who has become aware of an "ontological ecology"—that is, the transition from a reflexive conception of ontology (*cogito ergo sum*) to a relational one (*dialogo ergo sum*). This is an inclusive relation and must therefore be differentiated from the two traditional models, which are: (1) the exclusive dialectics: I am your opposite; (2) the relational predication: I am the link or means that binds me to someone else. An inclusion built on the epiphany of the encounter and on interpretation. One could say that the hybrid has become aware of its need for otherness and is humbled precisely for this reason. It is not anthropocentric because it stopped thinking in universal terms and therefore stopped searching for a centre. This is a very delicate step, but I do think that today's post-humanist philosophy offers the best tools to face the human challenges to come.

Glossary

Anthropocentrism From the Greek "*anthropos*" (human, human being) and "*kentron*" (centre). Anthropocentrism is a principle considering human beings as the centre of the world. I would like to define three forms of anthropocentrism: (1) ontological, (2) ethical and (3) epistemological. The ontological anthropocentrism is based on the idea that human being is autonomous in building his own identity and that the human being has particular value. The ethical anthropocentrism, which some believe corresponds to speciesism (see Speciesism), argues that human beings have no moral duties towards non-humans, except, at most, only that of well managing their exploitation. The epistemological anthropocentrism is based on the idea that human beings are the measure and universal dimension by which to describe and explain any non-human reality. Besides, there are two forms of anthropocentric tendency: (i) the anthropocentric point of view or naïve anthropocentrism, to be considered as a kind of human projection of himself in the world; (ii) the anthropocentric ideology, that, on the contrary, comes from a philosophical paradigm linked to post-sophism and humanism.

Anthropo-decentralization Act of moving away from the centre of our innate human characteristics by encountering a non-human otherness, thereby making a different existential dimension desirable. From my personal perspective, humans do not lack anything from the biological point of view in order to respond to their peculiar needs—as argued by Arnold Gehlen; instead, they perceive themselves as lacking because of desire. Whenever encountering an animal otherness, a human being projects herself onto it through empathy and desire; therefore, the encounter is never alien and indifferent, and becomes a kind of revelation (see Animal Epiphany). The encounter's epiphanic nature makes the human system open to external contributions; hence the feeling of deficiency. Herefore, the latter is not an a-priori and not related to an effective need, but is the outcome of the desiring projection. Anthropo-decentralization makes the system unstable and needy of external contributions; this need creates the feeling of lack. In other words, it is not a lack due to a poor human somatic equipment, but the result of a kind of "love at first sight" happening during the encounter with otherness, which paves the way to the desire to reach new existential dimensions.

R. Marchesini, *Over the Human*, Numanities - Arts and Humanities in Progress 4, DOI 10.1007/978-3-319-62581-2

Anthropo-metrics The use of morphologic and expressive coordinates to read and interpret the external reality. It is a concept well described by Leonardo da Vinci's Vitruvian Man. According to anthropo-metrics, the human being is a neutral identity, devoid of any specific function: usually, the human hand is given as example in this sense: it is, suitable for many different functions, as opposed to a hoof or a claw. Therefore it is taken as measure of world representing the universal. Anthropo-metrics is essential to epistemological anthropocentrism with humanistic roots.

Anthropomorphism (projective *vs.* critical) Projective anthropomorphism is the act of conferring human attributes to the other species. It is one of the most relevant epistemological mistakes when studying non-human animals. We can say that, throughout process of learning about another species, anthropomorphism is misleading whenever it brings about an absolute and projective trait. On the contrary, it is necessary whenever it is critical, using evolutionary continuity as a comparative term between human and other animals in order to understand the similarities they share on the basis of: (1) homologies: traits shared by hereditariness; (2) analogies: similarities due to adaptive convergences; (3) universals, namely the basic traits that are needed for life. Critical anthropomorphism is based on the concept that the non-human animal is interesting and can be understood precisely because it is not extraneous—let alone alien—to us.

Anthropo-plastics The attitude of adapting the world to our image and likeness in two ways: (a) by anthropo-shaping, that is, by attributing a human physiognomy to the world; (b) by anthropo-metrics (see Anthropo-metrics), that is, by applying to the world our peculiar ratio and projectivity, which becomes both the outcome and confirmation of anthropocentrism itself. This bias seems to contradict the previous one, but actually becomes another discriminating factor. In projection, nothing is considered if not human, therefore this bias is a kind of anthropocentrism that diminishes the importance of other species in two ways: either denying it any peculiarities; or turning any non-human entity into something getting closer to us.

Anthropo-poiesis From the Greek "*anthropos*" (man) and "*poiesis*" (creation). Anthropological term indicating the self-creation processes of the social and cultural human dimension. According to zooanthropology (see Zooanthropology), it represents the human becoming, that is, the result of a process of ontological hybridization with non-human otherness (see Ontological hybridization). At the ethnographic level, it is undeniable, for example, that birds-of-paradise are very important in New Guinea cultures, Gruiformes in Sub-Saharan African cultures, capercaillies and grouses in Ladin cultures or cetaceans in Polynesian cultures. According to zooanthropology, living close to a certain species means build a permanent dialogue; the latter inevitably produces an overlapping between the non-human predicates and the shaping of a cultural identity. In this regard, consider the example of Maasai people (Kenya) and their dance: it is shaped on the kinesthetic schemes of the grey crowned crane. These motions have entered the human predicates not in a projective way, but as the result of a zoomimesis (see Zoomimesis), which, first of all, means introjecting and assuming non-human predicates.

Anthroposphere In ecology, this is a part of the biosphere characterized by the presence of human beings. In the book, I will use this term with reference to the concept of anthropomorphism, as a result of the human misleading feeling of ontological self-sufficiency and, consequently, the idea of a human autarchy that contains and separates the human from the non-human—a kind of self-celebrating delusion. The mistake lies in considering, culture, technology, machines as just an emanation of human beings, so that immersion in the anthroposphere becomes the realization of a fully correlated and autonomous dimension: a kind of regression to the maternal womb.

Autopoiesis Term coined by the philosopher Humberto Maturana in 1980, from the Greek "*auto*" (oneself) and "*poiesis*" (creation). An autopoietic system permanently re-defines itself and feeds itself, reproducing from the inside. In the book, I will use autopoiesis to describe the human being as a self-demiurge, therefore autarchic and solipsistic in his self-definition, heroic in subjugating the world, able to constantly re-negotiate his existential *limes*, indifferent to anything surrounding him in relation to his ontological profile (see Ontogenesis).

Biocentrism The philosophical view that the centre of the world is not the human being, but bios, life. In this sense, biocentrism recognizes the polycentrism peculiar to animal-being. In the book, I will use "functional biocentrism", related to the concept of critical anthropomorphism (see Anthropomorphism), because it focuses on animal-being, and not species-belonging and tries and understand the species-specific predicate on the basis of the need of functional correspondence. Cats and dogs have different predicates (see Predicate), but they share some meta-predicates (see Meta-predicate) peculiar to all animals. Behavior responds to the same adaptive postulate as other functions, because this is the logic of animal-being: all animals are connected to reality and act consequently.

Biophilia Term coined by Edward Wilson in 1985. It indicates the esthetic appeal of the non-human animal, a charm that, according to the author, can be attuned with the phylogenetic expectations of the human being. It is an a priori aesthetics, which is a seemingly risky hypothesis, claiming the existence of an innate perceptive Gestalt and, what's more, a kind of taste orientation that pre-exist any subjective experience. Nevertheless, if we consider aesthetics as just one form of elicitation, then it falls within the group of key signals. According to Wilson, the animal shape, just as a key signal, would be felt as rewarding by human senses, eliciting astonishment and admiration. In other words, our species would evaluate the animal shape as aesthetically rewarding in an innate manner (*a priori*), being fascinated by it. In the book, in response to Wilson's idea, I will propose instead the theory of zootropy (see Zootropy).

Declination (see Predicate) Term I use to identify the individual's process of species tendencies to certain experiences and adaptations to the environment: it is what makes every non-human animal unique. Often, we think of innate endowments as a defined set of features that standardizes all the members of a given species from the point of view of expression. On the contrary, the innate is like a schema, a structure able to adapt according to the developing path chosen by the individual. If cat-being foresees the predatory motivation, a towards-something motion, how, when and where predatory behavior will be expressed depends on the biographic experience of the subject: in this regard, I would say that the predicate "predatory" is declined in a specific and peculiar way in a specific cat. The individual declination of predicates transform characters, which are not longer like essences, but like unlocked laboratories: material at the subject's disposal to build their own existential dimension as the protagonist of their existence.

Epiphany or epiphanic event Animal epiphany is an important event in human history, one that we can define as a revelation of a possible and desirable existential dimension whenever a human being encounters an animal otherness and its different life condition, such as a bird's flight, a fish swimming, a bear's expression of strength, or a wolf's howl. It is the moment when the encounter with a non-human animal produces a desirable condition, that is to say it shows to the human a lack in his own state. Only then can the human being see himself, think about his own humanity, see himself naked, without any particular predicates, weak in comparison to other species, lacking any weighty significance. Epiphany is a projection into otherness, and, above all, a becoming otherness producing a desire and a feeling of insufficiency. At the same time, epiphany is a possession, otherness being introjected, and human being appropriates predicates of the other through imagination. With an animal epiphany, man catches sight of new existential dimensions and inaugurates the Overman, able to imagine himself inside the zoomorphic projection. In so doing, the human being experiments a bird's flight or the strength of bull, and feels a distance between his own magmatic and metamorphic identity and the animal one, steady in its own predicates.

Epimelesis From the Greek "*epimeleomai*" (taking care of), the term indicates: (i) being involved with the needs expressed by otherness, with its asking for help, with its objective condition of exposure and vulnerability, being attracted by its pedomorphic traits of juvenile expressions (see Neoteny); (ii) putting in place some behaviors of rescue and assistance, care and nurturing, taking charge of the other's difficulties and expressing consistent responses of help and altruism (seen as the capacity to overshadow one's own interests), up to self-sacrifice. Accordingly, I will use et-epimelesis to define a request for care.

Epimetheanism In the book, I will use this term to indicate the attitude of considering non-human animals as entirely closed-off inside a behavioral determinism, as if they were machines activated by some automatisms (see

Reification). In this sense, the expressive trait of the animal is interpreted, according to the myth of Epimetheus, as a enclosure within a class that makes the individual unable to be the protagonist of her own existence, with no possibilities to choose freely. Therefore, the behavioral endowment is described as a wire moving the puppet-animal, a switch to turn functions on and off: these are the explanatory models considered when we talk of instincts and conditioning. Epimetheanism can be explained as a needed expression—similarly to the fall of an object—a full comprehension of the animal being in her function. In this regard, Epimetheanism is opposed to Prometheism, which is the claim of man's full self-determination.

Epimetheus (*vs.* Prometheus) Epimetheus, literally meaning "the one who engages in hindthought". In Plato's homonymous writing, the Sophist Protagoras says that human nature is the outcome of a dialectic action between Epimetheus and Prometheus. The story of Prometheus, generous wit humans, has been told in different manners by Hesiod and Aeschylus, but it is Plato who used this myth to come to important ontological conclusions. The story is well known: Epimetheus distributes different biological virtues among various animals, but, when man's turn comes, there are no performative gifts left and our species seems condemned to inadequacy. At this point, Prometheus ("the one who looks ahead") intervenes and offers humans fire and *techne*. In other words, Epimetheus condemns us to biological insufficiency, whereas Prometheus offers us another dimension, the cultural one: on one side, the latter compensates for the limits of human nature and, on the other side, it allows our species—the only one among all animals—to take a leap beyond *physis* and create the basis to separate itself from the others.

Heteromorphy It is the content of diversity of any non-human animal. Behavioral diversity, that is, the species dimension, has been highlighted by ethology, which, contrary to Watson's and Skinner's behaviorist interpretation o, posits that expressive diversity is not only the result of environmental determinations but is partially innate and partially inherited from maternal and social learning. The species dimension has a particular immersion into the world (*umwelt*) and a catalogue of peculiar behaviors (ethogram). In the literature, we can read how Humanism has taken off any heteromorphy—that is, species-specific dimension—from non-human animals, and how the human dimension was taken as universal, whereas animal otherness was simply seen as lacking or symmetrical to man.

Heterospecific Term indicating any individual not belonging to the *Homo sapiens* species. I will use this term as a synonym of non-human. The term has a particular meaning in defining not only the trait of diversity, but also the trait of relationality: indeed, there are relationships with conspecifics, for example inter-human ones, and relationships with heterospecifics, that is to say the ones between the human and the other species. According to zooanthropology, relationships with non-human animals are fundamental to build the human predicates.

Materning The adoption, nurturing and, often, breastfeeding of cubs from other species by *Homo sapiens* women. According to zooanthropology, the materning phenomenon, currently still present in many extra-European cultures, is explained with the epimeletic inclination (see Epimelesis) peculiar to our species; in zooanthropology, motherhood is the incipit of the domestication process. In fact, any domestication involves an expression of motherhood (cub breastfeeding) and weaning (mouth-to-mouth food transfer), two actions that obviously require an epimeletic attitude on part of the adoptant.

Mythopoesis Inclination to transform human history in a myth, to create a mythical vision involving a hostile world, few implacable enemies and a hero able to fight some overwhelming obstacles. In the book, I will underline how the myth of a Promethean origin of human predicates represents the basis on which to build a heroic vision of the human path, describing man as a kind of David fighting and winning, despite his weakness, over Goliath-nature. We can find this representation also in the iconography related to Saint George and the Dragon or in the tale of Minotaur. This anthropocentric tradition, aimed to exalt the human being, has to necessarily degrade the performative traits of our species—describing them as insufficient, inadequate and incomplete—in order to increase the heroic imagine and mythical dimension of the human.

Neoteny/Neotenic characteristics First of all, we have to distinguish pedomorphism from neoteny. The first is the set of morphological traits present in young mammals (for example, big eyes, round face, prominent forehead, chubby cheeks) that, acting as et-epimeletic targets, are able to trigger parental behaviors. In this regard, there is an interesting study by Konrad Lorenz about the fact that all mammals share the pedomorphic traits, so the cub of a certain species can arouse care and nurturing behaviors even in a different species. Neoteny is, instead, the persistence of the juvenile traits in the adult, that is, after sexual maturity. Generally, domestic animals have neotenic traits, such as a dog's bark and the flag-up tail positioning in a cat. According to a few authors, the human being also presents some neotenic characteristics.

Ontogenesis From the Greek "*ontos*" (being) and "*genesis*" (creation, development). Ontogenesis is the general construction of an individual's identity and self-perception as a biographical entity. Therefore, ontogenesis is influenced by the environment and the socio-cultural milieu where the individual expresses herself since her first developmental stages. In mammals, ontogenesis allows the individual to adapt to the peculiar condition where the juvenile will grow. Ontogenesis takes place in a particular moment of life called "evolutionary age" and is characterized by the actualization of some development processes, such as bonding and socialization. In the human being and a few other species, ontogenesis is also the creation and identification of one's self following some coordinates mainly rooted in culture.

Ontological hybridization A capital event in zooanthropology and post-human philosophy, by which the human condition is no longer seen as an exclusive consequence of the biological human traits—both through emanation and compensation—but as the result of a projection into otherness, the result of human desire towards the other's predicates, that is, the result of dreaming of owning the qualities of other species. This projection into the non-human animal (literally, animal becoming)—expressed by shamanism—turns the human dimension into an ontologically hybrid shape. This means that man, as a cultural and developmental condition open to multiple possibilities, goes beyond *Homo sapiens*; in other words, it is impossible to explain the human dimension and its predicates by just observing man as a biological entity, insisting on extracting human qualities from human traits. In hybridization, the interface becomes wider to include "other" perspectives, and this makes the human being less closed-off inside his dimension of adaptive specificity.

Ontopoiesis A sort of meta-ontogenesis, we could call it an ontogenesis inside ontogenesis, made possible by a population's collection of a few traditions that give a particular direction to the youth's behavioral development. The birth of cultural traditions is thus an event causing some processes that goes beyond the ontogenesis. Ontopoiesis is therefore the formation of new coordinates of development, able to change the behavioral setting of a population within a species. We should notice that the transformation of the ontogenetical space from a well defined reality (a defined pack of information to acquire) into a variable space, linked to local or temporary traditions, causes the individual's dependence on the cultural milieu. Ontopoiesis emerges whenever the traditional ontogenesis, based on individual experiences (declination of the innate heritage) and on a species-specific apprenticeship (maternal learning) expands to include a third coordinate: the cultural one, linked to a tradition of the population.

Phylogenesis From the Greek "*philos*" (class, species) and "*genesis*" (birth, creation, origin), it is the ramification of the lines of descent throughout life evolution. An outcome of phylogenesis is what I call species-specific dimension, that is, the creation of a particular position in the world, along with the refusal of any reference to objectivity or universalism to acquire some specific adaptive purposes.

Predicate/Meta-predicate The way in which the animal—with her species, individual or temporary characteristics—shows and expresses her needs. For example, every species has a different way to perceive, communicate, pro-act, re-act and so on, that is to say every species gives a certain predicative shape to the different chapters of existence. The term "meta-predicate" indicates the characteristic of a-specificity of animal being, that is, what it means to perceive, communicate, pro-act and re-act, to stick to the example above. The analysis of a meta-predicate indicates the way we think about animal being. The question "What does it mean to be an animal?" is the starting point of a philosophy of animality that aims to criticize the anthropomorphisation of the animal and the

Cartesian reduction of the animal to a "machine". Sentience, desire, subjectivity are meta-predicates, because they are detectable in any declination of animal being.

Prometheus (see Epimetheus)

Reification (*vs.* Anthropomorphisation) Equating a subject to an object (*res*). Zooanthropology assumes that the broad interpretations of non-human animals coming from reification and anthropomorphisation are wrong and misleading.

Syllegy From the Greek "*sin*" (togheter) and "*lego*" (to gather). Motivation characterizing our species and based on gathering, classifying, categorizing, inserting, that is, on the pleasure to build a collection able to put together uniformity and gradual diversification, using a logic of membership (put everything in the right place in a vision of order). According to zooanthropology, the syllegy related to *Homo sapiens* would have triggered the domestication process and, more in general, the phenomenon of hybridization between the human being and the other species.

Synmorphy In the book, the term means "sharing of the same shape".

Speciesism Term indicating the attribution of a different value and moral status to different individuals on the basis of the species they belong to. Speciesism does not consider inherence—what is specific to the subject—and bases moral value on what is shared: being different by definition, the non-human animal is marginalized from the condition of moral patient. Speciesism is therefore a tautology: the non-human has no rights *qua* non-human. Diversity cannot be the key aspect of this invalidation, because this would mean assuming the universal idea of interest, that is, in fact, a speciesism.

Technopoiesis In zooanthropology, it represents the effort to use some tools to externalize the qualities of the animals' bodies. The technopoietic event, like the animal epiphany, leads to some existential shifts and causes a decentralisation.

Technosphere We should not consider the technosphere like a womb that immerses man in a projective fractal of himself and cutting him off reality; on the contrary, it is an interface that increases the gradient of participation. In and through technology, man can diminish his anthropocentered contents (see Anthropodecentralisation), his gravitation around the species dogma, and leave the solitude of a species-specific perspective to be contaminated by otherness. Techne does not separate; instead, it connects. It has no narcissistic effects, as we might assume with the concept of a womb-technosphere, but promotes dialogue: techne is the widening of the human dialogic interface.

Threshold (effect) A focal point in the zooanthropological theory. The non-human animal represents a referential otherness for the human being: a threshold, a door that opens onto ecstatic experiences with a high degree of contamination. Animal otherness places man on the edge between the sphere of sharing, identification, belonging and the sphere of estrangement, astonishment, non-involvement, facilitating the processes of contamination, inspiration, orientation, and, at the same time, causing a few problems, questions and setbacks. The non-human animal is a threshold, an entity able to contaminate the human being adding new contents to his identity through models, inspirations, setbacks, and shifts of meaning.

Umwelt In the definition provided by von Uexküll in the late nineteenth century, *Umwelt* indicates the world-environment peculiar to every species, each characterised by a peculiar immersion in a world-context. This immersion is not only sensorial but is based on (1) motivations; (2) emotions; (3) knowledge; (4) specific cognitive functions. In the book, I have stressed how erroneous it is to consider the different Umwelten as separate monads, as these immersions present some overlapping areas. In other words, there are wide shared areas among different species, detectable through critical anthropomorphism and functional biocentrism.

Zooanthropology Discipline studying the relationships between the human being and the other species in the different dimensions of encounter and considering the referential value possibly arising therefrom. In zooanthropology, this is a special event, which should not be interpreted as usual, denying its traits of intersubjectivity and referentiality. In this sense, zooanthropology goes beyond (a) the traditional psychological approach and its vision of the animal a stimulus and a surrogate; (b) the instrumental vision—this animal is good for x—typical of the anthropological and zootechnical approach; (c) the interactive vision coming from the ethological or ecological views, which evaluates the relationship in terms of species and not of individuals. Zooanthropology was born around the 1990s, but it acquired a better defined epistemological structure thanks to my researches on the hybridization value of relations between the human being and the non-human animal in constructing human predicates.

Zoomimesis Transformation of the human being following an epiphany due to the encounter with the non-human animal. It should not be confused with the mere imitation of the non-human, because this would mean (a) considering any mimetic act as "external/appositive", imitation without interpretation; (b) considering the result as "zoomorphic-objective", unable to bring any trait of novelty or creativity. Zoomimesis opens the human being to some possibilities of anthropo-poiesis—by which human becoming is necessarily shaped by culture—that contradict the autarchic concept of culture. Zoomimesis is the result of a dialogical process, an interpretation that involves a man able to represent the animal connotations—like an actor interpreting a role using his own artistic personality—and not merely imitate them.

Zoomorpheme The morphological structure made available by the non-human animal and included into the process of referential thinking.

Zootropy Inclination typical to our species to look at other animals and recognise them as social counterparts, so as to include them in the anthropo-poietic processes. The theory of zootropy, which I introduced at the end of the 1990s, indicates a clear separation from other non- or proto-zooanthropological explanatory structures, not based on the dialogic/referential role of the non-human animal. The theory of zootropy causes an important epistemological shift, because it considers the heterospecific interesting not as stimulating (despite being a source of astonishment due to its phenomenal richness/variety, thanks to the role of key signals, able to elicit the different motivations, due to its mimetic potential as a recognisable living being), but as an interlocutor endowed with referentiality, able to bring new contents by means of direct influence.

Index

R. Marchesini, *Over the Human*, Numanities - Arts and Humanities in Progress 4, DOI 10.1007/978-3-319-62581-2

MIX
Papier aus verantwortungsvollen Quellen
Paper from responsible sources
FSC® C105338

If you have any concerns about our products,
you can contact us on
ProductSafety@springernature.com

In case Publisher is established outside the EU,
the EU authorized representative is:
Springer Nature Customer Service Center GmbH
Europaplatz 3, 69115 Heidelberg, Germany

Printed by Libri Plureos GmbH
in Hamburg, Germany